Relative Logic for Intelligence-Based Systems

Silviu Guiasu

York University
Canada

Advanced Knowledge International

Publisher and Distributor

Advanced Knowledge International Pty Ltd
PO Box 228
Magill, Adelaide
South Australia, S.A., 5072
Australia
info@innoknowledge.com

ISBN 0-86803-978-0

Printed in Australia

International Series on Advanced Intelligence

Series Editors:

R. J. Howlett
University of Brighton
R.J.Howlett@bton.ac.uk

L.C. Jain
University of South Australia
L.jain@unisa.edu.au

Advisory Board:

Professor Dr Robert Babuska
Delft University of Technology
The Netherlands

Professor Dr Ian Cloete
International University in Germany
Germany

Professor Dr Pierre Glorennec
INSA de Rennes
France

Professor Dr Chris Harris
University of Southampton
United Kingdom

Professor Dr Beatrice Lazzerini
University of Pisa
Italy

Professor Dr Robert Parkin
Loughborough University
United Kingdom

Professor Dr Takushi Tanaka
Fukuoka Institute of Technology
Japan

Contents

PREFACE

Somebody says: "Light consists of tiny corpuscles." Somebody else says: "Light is a continuous wave." The classic logic insists that only one of these sentences is true and the other one should be false. The relative logic says that both sentences may be true, depending on the context. If something is impossible today, the relative logic says that, eventually, it could become true, or possible, or still remain impossible, depending on time, context, and circumstaces. The problem with truth is its many varieties. The objective of this book is to discuss some main features of a relative logic which is probabilistic, temporal, circumstantial, and often subjective. As John von Neumann said: "Truth is much too complicated to allow anything but approximations."

Sometime, somewhere, a publisher said that each equation included in a book would halve the sales. Unfortunately, it is impossible to present the relative logic without using a mathematical formalism. On the other hand, logic is a topic of interest for a large category of people of different professions. In order to make a compromise, the first chapter of this book, the longest one, deals with considerations about the relative logic, meant for readers who have no mathematical background at all and, consequently, it uses very few elementary mathematical expressions. By contrast, the other four chapters heavily rely on a mathematical formalism which is, however, selfcontained and therefore easy to assimilate and follow.

The first chapter discusses the relativity of logic. Paradoxes, humour, ambiguity of the everyday language, mystery literature and movies, learning, and decision process illustrate the flexibility of our ways of reasoning and the contradictions that they generate. The presentation is loose and the readers are left to draw conclusions by themselves about why such contradictions do occur.

The next four chapters are shorter and technical. They deal with the mathematical formalism of the two-valued relative logic, multi-valued relative logic, and stochastic logic, followed by two applications to pattern-recognition and to the process of weighting credibilities in decision making. Instead of following an axiomatic approach, common in the presentations of mathematical logic, which starts from a minimum number of primitive definitions and rules, the method used in this monograph is to start from the definitions of all basic concepts and all rules of the corresponding relative logic, and show how they can be subsequently used for deriving some consequences. The treatment is not exhaustive but it is operational and functional; the readers can continue deriving other consequences by themselves.

An appendix refers to the problem of reaching a verdict by weighting evidence. None of the references mentioned at the end of the book deals explicitly with the relative logic as it is presented here but the titles listed there are related to the topics discussed here.

In the logic of every human being we can detect a variable amount, but never negligible, of lack of logic. Except speed, where we are certainly inferior, perhaps this is the main difference between the human and electronic intelligence. Sometimes, the humans have the tendency, call it folly or inspiration, of jumping out of their program, over their program, against their program, therefore against the current logic. There is something tragic and also sublime in this. The extremes meet.

I cannot end this preface without thanking Dr. Lakhmi C. Jain, Professor of Knowledge-Based Engineering, Director/Founder of KES Centre, University of South Australia in Adelaide, for his decisive role in writing and publishing this quite unusual book, with my admiration for his open mind, amazing productivity, and vision.

York University, Toronto, March 22, 2003

Chapter 1. THE RELATIVITY OF LOGIC

Long ago, George Boole was concerned about mathematically formalizing the classic logic. The next comments, reproduced a century later, testify about his ambitious goal. "That which renders logic possible, is the existence in our minds of general notions, our ability to conceive of a class, and to designate its individual members by a common name. The theory of logic is thus intimately connected with that of a language." (Boole, (1956, p.1857)). "The mathematics we have to construct are the mathematics of the human intellect." (Boole, (1956, p.1858)).

The content of symbolic logic does not differ from the content of ordinary logic, the discipline systematized by Aristotle and elaborated by medieval thinkers. Both symbolic and traditional logic are concerned with the general principles of reasoning and both make use of symbols. Thus, traditional logic uses the words of the respective language, whereas symbolic logic employs a specially devised set of marks called ideographs. The essential characteristic of symbolic logic, known also as mathematical logic, is generality. It tries to keep the ambiguity of the ordinary language at a minimum. Leibniz is credited with the first consistent use of symbolic logic. George Boole was the second founder of symbolic logic. A complete and workable calculus is achieved and operations of the mathematical type are systematically applied to logic. On the other side, we do not write poems in symbolic logic.

> **Aristotle** (384-322 B.C.) Greek philosopher, sometimes referred to as "the Stagirite". Born at Stagira in Thrace, he studied at Athenes under Plato, became tutor to Alexander the Great, and in 335 B.C. opened a school at Athens. When Alexander died, he was forced to flee to Chacis, where he died.

> **Gottfried Wilhelm Leibniz** (1646-1716). German philosopher and mathematician. Born in Leipzig. A versatile genius, he did a significant start in the project of creating a universal symbolic reasoning. This was to be a calculus which would cover all thought, and replace controversy by calculation.

George Boole (1814-1864). British mathematician. Self-educated. He was a professor at Queen's College in Cork, from 1849. He published *The Mathematical Analysis of Logic* in 1847, and the seminal book *An Investigation of the Laws of Thought on Which are Founded the Mathematical Theories of Logic and Probabilities* in 1854. Creator of the Boolean Algebra in symbolic logic, a mathematical tool for manipulating sentences in the general framework of the classic logic. As he mentioned at the beginning of his book on the *Laws of Thought*, "[Logic and Probability] instruct us concerning the mode in which language and number serve as instrumental aids to the process of reasoning... a science of the mind is therefore possible."

The foundation of modern mathematics is done using set theory created by Georg Cantor at the beginning of the 20th century. The operations with sets (union, intersection, complementarity, inclusion) are similar to the operations from classic logic (or, and, non, implication). The use of set theory has introduced more rigour in mathematical proofs but has also showed that mathematics is not free of paradoxes. It is not surprising, therefore, that the chapter on Cantor from the history of mathematics book written by E.T. Bell (1965, pp.555-580) is entitled "Paradise lost". An excellent, rigorous, and compact presentation of set theory may be found in N. Bourbaki (1968).

Georg Ferdinand Ludwig Philipp Cantor (1845-1918). Jewish mathematician. Born in St. Petersburg, Russia. Taught at the University of Halle, in Germany, from 1869 until 1905. He created the theory of aggregates (Mengenlehre), known today as the set theory, on which the modern mathematics is based.

Deducing correct inferences from given premises is not the only object of logic. It rather instructs us concerning the mode in which language serves as an instrumental aid to the process of reasoning. Fortunately or unfortunately, depending on how we look at it, the classic logic and its symbolic variants cannot encompass the richness of human reasoning. The following informal comments are to be taken only as a rambling and unsystematic illustration of the main topic of this book, namely, that logic is relative and, coping with uncertainty, essentially depends on circumstances, context, experience, and time. By contrast, the next chapters will insist on the systematic formalism of the relative logic as a generalization of the classic logic and its many extensions.

1.1. Logical paradoxes

Thinking correctly has always been a virtue whereas being illogical, a major insult. Logic is a science dealing with certain properties of human reasoning, an analysis of the structure of reasoning. The logician's main concern is with an analysis of the structure of correct reasoning. The rules of inference in a system of logic are simplifications of some aspects of the complex principles of actual human reasoning. Generally, from true premises, such rules allow the inference only of a true conclusion. The goal of classic logic is the pursuit and preservation of truth, in order to sort out the true statements from the others, which are false. In classic logic a proposition is either true or false but not both. We deal with a *paradox* when starting from apparently acceptable premises and reasoning in an apparently acceptable way we obtain an apparently unacceptable conclusion. It is generally difficult to decide whether the conclusion is however acceptable or the premises and reasoning contain flaws. They reveal a kind of thinking dilemma which normally bothers us. Many details may be found in Bourbaki (1999), Dieudonné (1992), Martin (1992), Paulos (1980, 1985, 1991), Sainsbury (1988), and Smullyan (1992).

The name paradox comes from the Greek *para* and *doxa*, meaning *beyond belief*. According to Webster's New Collegiate Dictionary, by paradox we understand a statement that is seemingly contradictory or opposed to common sense and yet is perhaps true or an argument that apparently derives self-contradictory conclusions by valid deduction from acceptable premises. Synonyms are antinomy and anomaly. Wherever met, a paradox is a source of surprise, a challenge to our aspiration towards a simple unique truth. But, as somebody put it, the problem with truth is its many varieties. Some attempts at solving paradoxes have resulted in new types of logic.

The liar paradox.

A classic paradox refers to Epimenides, the Cretan who made the immortal statement: "A Cretan is a liar." Let us denote by P this proposition. If P is true, then Epimenides, being a Cretan, does not tell the truth and therefore P is false. If P is false, this means that its negation "A Cretan is not a liar" is true. If Epimenides is such a Cretan who does not lie, then what he says, i.e., proposition P, is true. On the other hand, if P is false and Epimenides is a Cretan who does lie, then P is proven true because in classic logic, according to the law of double negation, non-non-P is just P.

A variant of the above paradox is "This statement is false." As mentioned before, in classic logic each statement is either true or false but not both. If the statement "This statement is false" is true, then it is false. Conversely, if it is false, then it is true. Contradiction!

The above argument uses the assumption that "This sentence is false" is either true or false. In order to deny this assumption, Bocivar proposed to deal with the Liar Paradox by adopting a three-valued logic in which the third value, paradoxical, is to be taken by such trouble making sentences. But, in such a case, the sentence "This sentence is either paradoxical or false" is paradoxical or false if true, true if paradoxical, and true if false.

Tarski tried to solve the Liar Paradox by introducing the hierarchy of languages. According to him we have to distinguish: Level 0: The object language O; Level 1: The metalanguage M_1, which contains means of referring to expressions of O and the predicates "true-in-O" and "false-in-O"; Level 2: The meta-metalanguage M_2, which contains means of referring to expressions of M_1 and the predicates "true-in-M_1" and "false-in-M_1"; etc. In this hierarchy of languages, truth for a given level is always expressed by a predicate of the next level. Consequently, the sentence "This sentence is false-in-O" is itself a sentence of M_1, and hence cannot be true-in-O and is simply false and not paradoxical.

Buridan's paradox.

Another variant of Epimenides' paradox is Buridan's sophism. According to it, when Socrate in Troy says "What Plato is now saying in Athens is false," Plato in Athens says "What Socrate is now saying in Troy is false."

The regiment's barber paradox.

Sometime, somewhere, there is a regiment in a fortress and only one barber. According to a decree issued by the general in command there, nobody is allowed to have a beard and the barber shaves all and only those who do not shave themselves. What happens with the barber himself? Who shaves the barber? He is not allowed to have a beard and to shave himself. The barber shaves himself if and only if he does not. The rigid rules imposed by the general in command generate a contradiction.

Russell's paradox.

A set is a collection of entities called elements. Often, a set is characterized by a certain property that must be satisfied by the elements of the corresponding set. Thus, the set of even positive integers not exceeding 8 contains four elements {2,4,6,8}. A set may contain a finite number of elements, like the set of the books in my house, or an infinite number of elements, like the set of odd positive integers. If a is an element of the set A, the sentence "the element a belongs to the set A" is denoted by $a \in A$, where $\in$ is the symbol denoting "belongs to". There are sets that do not belong to themselves or are not elements of themselves. Thus, a set of apples is not an apple itself. But there are also sets that are elements of themselves. Take the set of nonapples. Anything, concret or abstract, thing, concept, or living creature, that is not an apple is an element of this huge set. A baseball bat, a child, a grape, a town, a song, a poem, an idea, are all elements of the set of nonapples. Obviously, such a set is not an apple and therefore belongs to the set of nonapples, i.e., is an element of itself. Russell's paradox refers to the set R of all sets that are not elements of themselves, i.e., the set whose elements are the sets that are not elements of themselves. The necessary and sufficient condition for something to belong to R is that it be a set that is not an element of itself. We should like to know what kind of set is R itself. Is the set R an element of itself or not? Let us analyse the two possibilities:

(a) If R is an element of itself, then it must satisfy the property that defines the set R, i.e., it cannot be an element of itself. Therefore, if R is an element of itself then it is not an element of itself. Symbolically, we write: $(R \in R) \Rightarrow (R \notin R)$, where $\Rightarrow$ means "implies" and $\notin$ "does not belong to".

(b) If R is not an element of itself, then it satisfies the property that defines the set R itself and, therefore, must belong to the set R. Thus, if R is not an element of itself, then it is an element of itself. Symbolically, $(R \notin R) \Rightarrow (R \in R)$.

Summing up, R is an element of itself if and only if it does not belong to itself. Symbolically, $(R \in R) \Leftrightarrow (R \notin R)$, where $\Leftrightarrow$ means "equivalent". The set R is contradictory.

In order to solve such a paradox, Russell introduced the hierarchy of types. The entities are placed at different levels of a hierarchy: Type 0: Individuals; Type 1: Sets of individuals; Type 2: Sets of sets of individuals; etc. A relation of the form $R \in S$ has sense only when R is of type k and S of type $(k+1)$. If r is an individual, R a set, and $\mathcal{R}$ a set of sets, then such relations as $r \in R$ and $R \in \mathcal{R}$ are valid but $r \in r$ and $R \in R$ are not. Russell's paradox could be avoided by using the so called "Poincaré's vicious circle principle", according to which whatever involves all of a collection must not be one of the collection.

> **Bertrand Arthur William Russell** (1872-1970). British philosopher and mathematician. He taught at Trinity College in Cambridge, England, and in several universities in USA. **Alfred North Whitehead** (1861-1947). British philosopher and mathematician. He was a professor of applied mathematics in London, between 1914-1924, and a professor of philosophy at Harvard, between 1924-1937. Russell and Whitehead, both at Cambridge at that time, published three volumes of *Principia Mathematica* in 1910-1913, an attempt to construct the whole of mathematics by logical deduction from a small number of concepts and principles.

Cantor's paradox.

No set can be larger than the set of all sets, but, for any set, there is another, the set of all its subsets, which is larger than it is. The paradox resulting from Cantor's set theory formulated by Bertrand Russell in 1903 may be stated in a brief form. Let S be the set of all sets that are not members of themselves. Question: is S a member of itself? If so, then it is not a member of itself; if not so, then it is a member of itself. This is similar to the ancient paradox of the Cretan who said that all Cretans lie.

Self-referential paradox.

There are many variants of this paradox. Basically, they are all based on the following scheme: Assume that X can do anything. Can X make a stone so heavy that X cannot lift it? X can because X can do anything. But if X could make it, there would be something X could not do, i.e., lift it.

The postcard paradox.

On one side of a postcard is written: "The sentence on the other side of this postcard is false." On the other side of the same postcard we read: "The sentence on the other side of this postcard is true." A different form of the same paradox refers to two consecutive sentences: "The next sentence is false. The previous sentence is true." The contradiction comes again from stating that a certain proposition is both true and false, which in classical logic is not allowed. But let us mention an anecdote involving a somewhat similar construction: During a reception, Voltaire made the following comment: "Marquis X says that I am stupid. I think that he is intelligent but, perhaps, we are both wrong." As the construction here is not self-referential, there is no contradiction but an ellegant, indirect way of denying.

The book paradox.

An author writes a mathematical monograph and its preface consists of only one statement "This monograph contains at least one false sentence." Then, for sure, the monograph contains at least one false statement because if there is no false statement outside the preface, then the preface itself is false. A well-known mathematician wrote an entire book, excellent in fact, about Bertrand Riemann's eight-page famous paper on the distribution of primes. Just from the beginning, he strongly advises the reader to read only the original version of the classical works and not the second-hand or third-hand sources. It was like saying "Do not read my book!"

Get what you want.

Here is an easy way of forcing a yes answer to the second question, whatever the answer to the first question is:
(1) Will you answer this question in the same way that you will answer the second?
(2) Will you marry me?

Protagoras versus Euathlus.

Euathlus wanted to become a lawyer and Protagoras offered to teach him rethoric. Euathlus paid half of the tuition fees and as far as the second half of the payment was concerned the following agreement was reached: Euathlus would pay the second half after winning his first case in court. As Euathlus delayed to start his practice, Protagoras decided to ask for money but the former pupil refused to pay. The case is brought to court and the following dialog takes place:

Protagoras: "Euathlus maintains that he should not pay me. If he wins this case, this would be his first case won and therefore he ought to pay me according to our agreement. If he loses his case, he ought to pay me by the decision of the court in my favour. Consequently, whatever happens in court, he must pay me."

Euathlus: "Protagoras maintains that I should pay him. If he wins this case, I will not win my first case in court and, according to our agreement, I do not need to pay him. If he loses his case, I do not have to pay him by the judgement of the court. Consequently, whatever happens in court, I do not have to pay him."

Identifying inhabitants.

We have a land in which every inhabitant either always tells the truth (a T-inhabitant) or always lies (an F-inhabitant). There are two minorities on this land, the P-inhabitants and R-inhabitants, respectively. A P-inhabitant always tells the truth and, therefore, is a T-inhabitant. An R-inhabitant always lies and, therefore, is an F-inhabitant. What statement could an inhabitant make that would convince a visitor that she always tells the truth but is not a P-inhabitant? The answer is that it would be enough for her to make only one statement: "I am not a P-inhabitant." Indeed, an F-inhabitant, always lying, could not make such a true statement because indeed only truth-tellers are P-inhabitants. Therefore she must be truthful, which means that she is a truth-teller but not a P-inhabitant. How should an inhabitant of that land introduce herself in order to be identified, without any ambiguity, as: (a) a P-inhabitant; (b) an R-inhabitant; (c) an F-inhabitant who, however, is not an R-inhabitant?

The example just given is something more than a simple game of mind. It reflects a deeply troubling dilemma if we replace land by *system of state-*

ments, *T*-inhabitant by *true statement*, *F*-inhabitant by *false statement*, *P*-inhabitant by *provable statement*, and *R*-inhabitant by *refutable statement*. The logicians Gödel and Tarski have proved that there are important domains where there are true sentences that are not provable inside the system, like the *T*-inhabitant who is not a *P*-inhabitant from the above story.

Gödel's idea was to translate this old logical paradox into mathematics itself, more exactly in number theory, the most clean-cut chapter of mathematics. He constructed a code, known as "Gödel numbering", assigning numbers to symbols and sequences of symbols. Each statement of number theory, being a sequence of symbols, gets a Gödel number, i.e., a code word consisting of some digits like a student identity number, a label, or social insurance number. Gödel showed that there is a Gödel number, obtained by taking a digit from each of the Gödel numbers assigned to a set of statements of number theory, by a kind of diagonal process, that corresponds to a non-provable sentence in number theory. Details may be found in Nagel and Newman (1958), Péter (1961), and Smullyan (1992).

> **John von Neumann** about **Gödel**: "Gödel was the first man to demonstrate that certain mathematical theorems can neither be proved nor disproved with the accepted, rigorous methods of mathematics. In other words, he demonstrated the existence of undecidable mathematical propositions. The result is remarkable in its quasiparadoxical 'self-denial': It will never be possible to acquire with mathematical means the certainty that mathematics does not contain contradictions. This is not a philosophical principle or a plausible intellectual attitude, but the result of a rigorous mathematical proof. Gödel actually proved this theorem, not with respect to mathematics only, but for all systems which permit a formalization, that is a rigorous and exhaustive description, in terms of modern logic: For no such system can its freedom from inner contradiction be demonstrated with the means of the system itself." ("Tribute to Dr. Gödel," March 1951, von Neumann archives, Library of Congress, Washington, D.C.)

Zeno's paradox.

A race is arranged between Achilles and a tortoise. As Achilles is a much faster runner, the tortoise is given a head start. Starting the race, Achilles has to get first to the place from which the tortoise started. But, in the meantime, the tortoise has advanced a little farther. At the second stage of his efforts for making up the handicap, Achilles has to get to the new place the tortoise is at. But, in the meantime, the tortoise will have advanced a little farther again. And so on. As long as the tortoise keeps going, Achilles can never catch it.

1.2. Ambiguity and dialectic thinking

The everyday language may be used in a very ambiguous or even sense-less way. When a sentence may be interpreted from different viewpoints, or contexts, it generates ambiguity. Thus, a candidate who applied for a new academic position, just advertized, is informed by a member of the re-spective department that: "I got the impression that the announcement has been taylored for somebody, but you have very strong chances." Was the an-nouncement taylored in order to satisfy some requirements imposed by the administration or to fit the position to the qualifications of a specific candi-date? The biggest tragedy ever involving commercial airplanes happened at the airport in Palma de Mallorca, a couple of years ago, when two full Boeing 747 jumbo jets collided before taking off, due to a language misunderstanding during the conversation between the two pilots and the control tower.

The entire literature on absurd is based on this possible ambiguity and on how senseless the language can be. To give only one such example: "I sleep very late. I commit suicide at 65%. My life is very cheap, it's only 30% of life for me. My life has 30% of life. It lacks arms, strings and a few buttons. 5% is devoted to a state of semi-lucid stupor accompanied by anaemic crackling. This 5% is called DADA. So life is cheap. Death is a bit more expensive. But life is charming and death is equally charming." (Tzara, (1984, p.49)). Some sentences have some kind of sense but the context is incoherent. The spoken language could be even more ambiguous. Many words belonging to the basic vocabulary have the same or very close pronunciation as other completely different words (meet-meat, bitch-beach-beech, etc.) and their meaning depends only on the context.

Lewis Carroll's Alice and her encounters with all sorts of weird characters and situations have fascinated generations about the possibility of having strange worlds, very different from the one we are leaving our everyday life in. For me, on the funny long street International Drive in Orlando, Florida, the most impressive was the monumental house built upside down, challenging all architecture's rules. The theatre of the absurd wants to show that often human beings live in an irrational and meaningless universe without even being aware of.

Sometimes, the comic is generated by trying to be as logical as possible (if you do not know the answer, at least rely on common sense), or by "logically" referring to something apparently similar but from a different context, or by unnecessarily inserting some personal, otherwise innocent, remarks in the wrong place at the wrong time. Here are some children's views on music collected by a music teacher (Harold Dunn) from Missouri:

"Refrain means don't do it. A refrain in music is the part you better not try to sing."

"A virtuoso is a musician with real high morals."

"A contra-bassoon is like a bassoon, only more so."

"I know what a sextet is, but I had rather not say."

"Just about any animal skin can be stretched over a frame to make a pleasant sound once the animal is removed."

"When electric currents go through them, guitars start making sounds. So would anybody."

"Henry Purcell is a well known composer few people have ever heard of."

"Tubas are a bit too much."

"Most authorities agree that music of antiquity was written long ago."

On the other hand, the language allows to introduce a meaning, a sense, a significance, a nuance, a flavour, or a surprise in our way of using words and sentences. This is why we smile when Woody Allen, in the movie *Sleeper*, objects to the intended reprogramming of his brain by saying: "You can't touch my brain, it's my second-favorite organ!"

Whether we admit it or not, the process of drawing conclusions during a logical analysis is heavily dependent on many factors like culture, education, ideology, interests, and facts from present or past experience. To simplify, we call these factors circumstances. Our conclusions are based not only on the rules of logic but on some circumstances taken into account as well. When future events do not confirm our logical predictions, the surprise is induced by the fact that some essential possible circumstances have been ignored. A clever author of a good detective story will do his best for highlighting certain secondary circumstances while keeping other essential circumstances in obscurity, in order to make the reader follow a wrong or roundabout path in his logical pursuit for the discovery of truth. After reading hundreads and hundreads of detective novels, where normally the culprit is some character from the respective story, based on this past reading experience, it is very surprising to forsee that the murderer could be just the narrator of the entire story or all the characters involved in the story, as it happens in Agatha Christie's *The Murder of Roger Ackroyd* and *Murder on the Orient Express*, respectively. When the available evidence overwhelmingly incriminates the only suspect, a very popular character, but the jury is unanimoulsy biased in favour of the accused and the judge is too coward, the unexpected "not guilty" verdict is reached. Even beyond the subjective logic, the truth or falsity of different sentences essentially depends on some circumstances taken into account explicitly or implicitly, consciously or unconsciously. As Niels Bohr noticed: "There are trivial truths and great truths. The opposite of a trivial truth is plainly false. The opposite of a great truth may be another great truth." The world as a whole and each of its parts have a dual nature (action and reaction, wave and particle, light and darkness, good and evil, heat and cold, male and female, the passive and the active, resistance and generation, negative and positive). Under some circumstances one side of this duality shows up while other circumstances reveal its coexistent dual side or allow the transformation of one side into its dual. The yin and the yang of the Oriental Taoism, or Hegel's dialectic of opposites have dealt with this duality. Moreover, under a magnifying glass, this duality proves to be only the tip of the iceberg represented by the plurality of the world and its parts. Not light or darkness but different nuances of light or darkness, not heat or cold but different degrees of temperature, not black or white but different levels of colour, not true or false but several possible logical values. The problem with truth is its many varieties.

Here are some key words for the different types of logics of interest: classic logic, multi-valued logic, diffuse logic, stochastic logic, global logic, fuzzy logic, deontic logic, deviant logic, quantum mechanics logic, dialectic logic, probabilistic logic, temporal logic, subjective logic, modal logic, dynamic logic, and now the relative logic.

And here are some disparate comments on the necessity of a more flexible logic, found in different lecture notes taken by me long ago:

T.H. Huxley: "It is the customary fate of new truths to begin as heresies and to end as superstitions."

From an old Italian movie: "Discipline is the logic of soldier."

G.C. Moisil: "Il y a un monde de nuances qui parmi ses 'oui' et ses 'non' a tant de 'peut-être' ... La pensée humaine est imprécise, donc la logique naturelle est beaucoup plus nuancée que la logique booléene."

A. Kaufmann: "Pour inventer il faut être subjectif et imprécis."

C. Peirce: "All things swim in continua."

L. Zadeh: "Informally, by approximate or, equivalently, fuzzy reasoning we mean the process or processes by which a possibly imprecise conclusion is deduced from a collection of imprecise premises. Such reasoning ... falls outside the domain of applicability of classic logic." "What is possible may not be probable; what is improbable need not be impossible." "Nature writes with a spray can rather than a ballpoint pen."

D. Dubois and H. Prade: "Probability theory seems to offer too normative a framework to take account of uncertain judgment." "What is probable is certainly possible; what is inevitable (necessary) is certainly probable."

About some technical terms of interest for a flexible, relative logic, Webster's New Collegiate Dictionary tells us that:

certain: sure; proved to be true; inevitable; assured in mind or action.

truc: just; truthful; being in accordance with the actual state of affairs; conformable to an essential reality; legitimate; rightful; conformable to a standard or pattern.

truth: a judgment, proposition, or idea that is true or accepted as true; the property of being in accord with fact or reality; veracity; verity.

probable: supported by evidence strong enough to establish presumption but not proof (a probable hypothesis); establishing a probability (probable evidence); likely to be or become true or real (probable events); likely; physical: statistical experiments, frequency of occurrence; epistemic: subjective judgment.

possible: being within the limits of ability, capacity, or realization; being what may be done or may occur; being something that may or may not occur (it is possible but not probable that he will win).

necessary: of an inevitable nature; inescapable; logically unavoidable; that cannot be denied without contradiction; compulsory; absolutely needed, required.

possible-necessary: Aristotle stressed their duality (if an event is necessary then its contrary is impossible).

plausible: superficially fair, reasonable, or valuable but often specious.

belief: a state or habit of mind in which trust or confidence is placed in some person or thing.

credible-plausible: anything that can be deduced from the corpus of knowledge is *credible*; anything that does not contradict the corpus is *plausible* (inductive aspect). *Likely* is near to *probable*; *plausible* is close to *conceivable*.

fuzziness means *vagueness*.

ambiguity refers to several possible contents or reference sets.

generality is a form of imprecision related to the process of abstraction.

informal principle: the degree of possibility of an event is greater than or equal to its degree of probability, which must be itself greater than or equal to its degree of necessity.

The stochastic logic is a nonstandard logic. It is not a rival of classic logic but rather a generalization of both classic and fuzzy logics. Its main objective is to take interdependence or global connection into account. It is both probabilistic and globally dependent of its context. The stochastic logic is a supplementary logic in the sense that it is compatible both to the classic logic which is nonprobabilistic, or strictly deterministic, and the fuzzy logic which may be viewed as a probabilistic and locally independent logic. It also incorporates the temporal logic as well. The probabilistic feature of the proposed logic is motivated by the necessity of coping with uncertainty in dealing with real life problems. Such uncertainty could have objective or subjective causes. Sometimes, the qualifications for being something are imprecise. Sometimes, the qualifications for being something are precise, but there is difficulty in determining whether certain subjects satisfy them.

Vague sentences may be neither true nor false. Apparently, Aristotle concluded that not every future tense sentence is true or false. Post (1921) took into account partial truth, by dealing with the truth values: wholly true, half true, and wholly false. B. Russell (1903) assumed that vagueness shows that classic logic is not applicable to ordinary language; this does not mean that it is false but simply inappropriate. J. Łukasiewicz (1930) introduced a third logical value, representing "the possible", denoted by $1/2$, different from 0, or falsity, and 1, or truth. He wrote "My presence in Warsaw ... at noon on 21 December of next year, can at the present time be neither true nor false. It is possible, but not necessary."

We have less confidence in precise statements and more confidence in vague statements. Vagueness makes an assertion compatible with more observed facts. Often precision may be increased to the detriment of credibility.

Probabilistic logic is a form of modal logic, proposed by Hans Reichenbach, in which true and false are replaced by probabilities ranging from 1 (true) to 0 (false). In fuzzy logic, as in probabilistic logic, variables can take any real values from 0 to 1 inclusive, but instead of representing probabilities that a sentence may be true, such as "The next toss of this coin will be tail," they represent the degree to which a variable is in a fuzzy set. Traditional nonfuzzy sets, called Cantor sets, or "crisp sets" are sets vith a value of 1. As L. Zadeh points out, if a concept is extremely fuzzy it becomes "vague". Thus "Mary will be back in a couple of weeks," is a fuzzy statement, but "Mary will be back sometime," is vague. The conclusions of fuzzy logic are fuzzy.

Instead of making vague arguments precise, fuzzy logic makes logic itself imprecise. True is defined as a specified fuzzy subset of the set of degrees, or values of truth (true, false, not true, very true, not very true, more or less true, rather true, not very true, not very false). Shaggy logic underlies both ordinary language and most human thinking. It is what gives human reasoning its superior power: the ability to play chess, to read poor handwriting, to write a summary of a novel, to play tennis, or park a car. The ability to manipulate fuzzy concepts is superior to that of any computer program yet devised. Heuristic reasoning is an informal reasoning by procedures that are not formally perfect but which resemble the intuitive way in which human mind actually works when confronted by a problem or when it has to cope with complexity. How the human mind works remains a profound mystery. Nerve-nets in the brain have a superficial resemblance to computer networks, and synapses can be looked upon as switching devices. So far no one knows even how memories are recorded, let alone how the brain solves difficult problems, invents a good scientific theory, or creates a memorable tune. The brain seems to have evolved so that its thinking involves a great deal of redundancy that protects it from transmission errors. The truth is that nobody today has a deep understanding of the kind of logic the human brain uses or how it manipulates it. No one even knows how an ant's brain processes its inputs and tells the ant what to do.

Modal logic deals with the concepts of necessity and possibility. A *nec-*

essary truth is one which could not be otherwise; a *contingent* truth is one which could be otherwise. The negation of an necessary truth is impossible or contradictory; the negation of a contingent truth is possible or consistent. Petrus Hispanus in *Summulae Logicales* has considered four modes of propositions: possible, contingent, impossible, and necessary. Aristotle in *Analytica priora* considered only three: real, possible, and necessary. I. Kant did the same. For Aristotle, possible does not mean all that can exist but rather what is opposed to necessary, i.e., what can and cannot exist. The complement of possible is possible. C.I. Lewis considered six modes (or modalities): true, false, possible, impossible, possibly false, necessary. The new notation, in which each operation is placed to the left of its operands in the symbolic multi-valued logic, was invented by the Polish logician Jan Łukasiewicz and is called as Polish notation. Thus, J. Lukasiewicz introduced the symbols: Np for "proposition p is false," Mp for "proposition p is possible," (with M from the German möglich which means possible), NMp for "p is not possible," (impossible), MNp for "non-p is possible," (contingens), $NMNp$ for "non-p is not possible," (necessarium).

H. Reichenbach believes that the logical concept of probability should replace the two truth values (true, false) from the classic logic. R. Carnap believes that this world shows a lack of distinction between "true", on one side, and "known to be true", "absolutely sure", "completely verified", "fully confirmed", or "having a probability 1", on the other side. The probabilistic concept of truth is not always referring to evidence. The probabilistic logic is rather an inductive logic.

At a macroscopic (coarse) scale, the classic logic is a good approximation of the stochastic logic. At the microscopic scale, fuzzy logic is locally a good approximation for the stochastic logic because it may be viewed as a kind of probabilistic logic which, however, assumes independence. At the atomic scale or at the megascale (the universe as a whole), where global connection (interdependence) is essential and the joint probability is the only concept allowing us to cope with complexity, the stochastic logic is indispensable.

The logical paradoxes have cast doubts about the validity of classic logic and the *tertium non datur* principle, i.e., the principle of excluded third. In any modal logic the distinction "true-false" from the standard logic is replaced by the larger distinction "possible-impossible". In the multi-valued logic, the basic idea is that propositions could have some other values than true and false, for instance the value "nonsense".

Pleading in favour of a probabilistic logic, H. Reichenbach (1949) gives the example of a person shooting at a target. Take the sentence "I'll hit the center of the target". After the shooting we measure the distance r between the point hit by the bullet and the centre of the target. The number $1/(1+r)$ is a measure of the degree of truth of the above statement. In this example there is only one person involved. If there are several, the performance of one shooter may essentially be influenced by the behaviour of the other ones.

Both "Life is short" and "Life is long" may be true, depending on the context. In the dialectic logic, which may be better called the "river logic", we have a conditioning by two types of past, present, and future events: from the river and from its borders. In an excellent book, J. Gaarder (1996) gave a brief but deep characterization of some important topics of the dialectic logic created, but obscurely explained, by Hegel (1843). The "world spirit", or "world reason", is the sum of human life, human thought, and human culture. Truth is subjective. All knowledge is human knowledge and there is no truth above or beyond human reason. Hegel did not believe that it was possible to make pronouncements on the eternal, timeless factor of human knowledge. His dialectic is a method for thinking productively, based on understanding the progress of history. There are no eternal truths, no timeless reason. The only fixed point is history itself, which is like a running river. A river is in a constant state of change. This does not mean that we cannot talk about it. How Gaarder put it: "You cannot say at which place in the valley the river is the 'truest' river. Every tiny movement in the water at a given spot in the river is determined by the falls and eddies in the water higher upstream. But these movements are determined too by the rocks and bends in the river at the point where we are observing it." The history of thought is like this river. Past tradition and material conditions at the time influence our way of thinking. We cannot claim that any particular thought is correct for ever. It can be correct, however, from where we stand. Things can be right or wrong in relation to a certain historical context. Reason is a dynamic process and the truth is this same process because there are no criteria other than the hystorical process itself that can determine what is the most true or the most reasonable. The reason is progressive when something new is being added. Like the river that becomes broader and deeper during its flowing process, the world spirit is developing towards an ever-expanding knowledge of itself. Truth is a process. When, on the basis of other previously proposed thoughts, a new thought, called thesis, is proposed, its negation, called antithesis, shows up as well. The contradiction and tension between thesis and antithesis is resolved later by the negation of the negation that induces the synthesis, which tries to accommodate the best of both points of view. Hegel's negative thinking tries to find flaws in an argument but, when such flaws are found in the respective argument, it preserves, however, the best of it. Thus, the tension between the concepts of "being" and "not being" is resolved in the concept of "becoming". As mentioned previously, the Danish physicist Niels Bohr is credited for saying that there are superficial truths, the opposite of which are obviously wrong, and profound truths, whose opposites are equally right. In classic logic time is frozen and there is no transition, development, and transformation of the concepts. Hegel's dialectic is a dynamic logic which tries to deal with opposites because reality is full of opposites: existence and non-existence, light and dark, attraction and repulsion, reason and intuition, male and female, positive and negative, physical and mental, attraction and repulsion, action and meditation, outbreathing and imbreathing, awake and asleep. Creation can only take place in situations of interactions of opposites, but all contraries cease to exist as such when they are viewed from a higher level than the one where their opposition takes place, allowing the acceptance of the whole with its negative and positive aspects. Also, the opposites may change places on different levels. Thesis gives rise to antithesis and their interaction gives synthesis.

> **Georg Wilhelm Friedrich Hegel** (1770-1831). German philosopher. Born at Stuttgart, he lectured at Jena and Nüremberg, and was a professor of philosophy at Heidelberg and Berlin. His writings include *The Phenomenology of Spirit*, *Logic*, and *Encyclopaedia of the Phylosophical Sciences*. According to him, mind and nature are two abstractions of one indivisible whole. The laws of thinking are in principle the laws of reality and logic must reflect the contradictions within nature. Each system by its own development brings about its opposite (antithesis), and finally a higher synthesis unifies and embodies both.

Elements of dialectic logic may be found in the old Taoist doctrine. The basic aim of the Taoist is the attaining of balance and harmony between the *yin* and *yang*, known as the two great powers, the two poles. This balance and harmony must be achieved both in one's self and in the world until the two are resolved into the One (J.C. Cooper, (1981)). Lao Tzu's *Tao Te Ching*, dated around 600 B.C. and said to have been more translated then any other book except the Bible, consists of about 5,000 words only. Taoism is taught by example. For the Taoist, virtue lies not in morality but in an inward quality of obedience to the Natural, in simplicity and spontaneity. Taoism frequently teaches through paradox and it is one of its paradoxes that simplicity is required to deal with complexity but simplicity is extremely difficult whereas it is easy to complicate things. The opposites can be seen as such in their relativity and their contrary aspects, but also in their unity. Not for nothing is the mean called "the happy mean" or medium. The objective consists in avoiding all extremes and establishing harmony. In a balanced person intellect and emotion, body and spirit, interact together in harmony. The One and the Many can never be separated since neither has any meaning except in relationship with the other. There is no "this" separate from "that". As mentioned in Jantsch (1981), according to Lao Tzu: "The hidden and the manifest give birth to each other," and acording to Chuang Tzu, the poet-philosopher of Taoism: "He who wants to have right without wrong, order without disorder, does not understand the principles of heaven and earth; he does not know how things hang together."

Taoism adopts a cosmological viewpoint as a religion. The creative power is looked for in Nature, not in some outside force standing separate from the things it creates. Yin, the physical, emotional, cerebral, inertia, the passive, the resistance, and yang, the intelligence, energy, the spiritual, the active, the generation, are the two great powers, the alternating forms of the creative force and are kept in proportion to each other. The opposites have a vital need of each other. As J.C. Cooper (1981) describes the dialectic relationship between Taoism's yin and yang: "In one sense the contraries are complementary and cooperative, in another they are mutually destructive or exclusive just as light and darkness cannot exist without eliminating each other, but the existence of each is only possible in juxtaposition to the other. There is a two-way traffic of similarity and dissimilarity, there are complementary qualities but also tension of balance and not of antagonism." In Taoism the cyclic view of life is basic: life gives way to death and death gives rise to new life. A cyclic time is self-contained, whereas in the Western culture a linear time has been preferred, except the ancient Greeks who basically followed the Indian cyclic tradition with Pythagoras pleading in favour of the eternal recurrence. A linear flow of time rises troubling dilemmas on the beginning and the end of the universe. A cyclic time does not assume a beginning and an end but a permanent return in an infinite endless process. A universal time is an illusion. There is no Swiss watch of the entire universe. Time is local and relative. Time ultimately has meaning only as correlations between events.

Taoism is not the only old doctrine containing early dialectic ideas. Buddhism is also credited as stating that if we want the truth to stand clear before us then we can never be "for" or "against", the struggle between "for" and "against" being considered as the "mind's worst disease", as mentioned by J.C. Cooper (1981).

The present book deals with a logic which allows more than two truth values, as in the modal and multi-valued logic, uses a probability distribution for describing the levels of certainty of its conclusions, as in the fuzzy and probabilistic logics, and essentially takes into account time, as in the temporal logic, the interdependence (connection) between components, as in the stochastic logic, and the specific context under which the reasoning is performed. Such a logic will be called *relative or dependent logic*. In classic logic the opposites cannot coexist: either A or non-A is true, but not both. The dependent logic says that both A and non-A may coexist, depending on the context. Dialectic understood in a narrow sense also does not accept that opposites could exist. There is a struggle between them followed by the unification thought like a melting into a new synthesis. In order to accept the peaceful coexistence between opposites, a relativization of logic is needed, where both p and non-p may be true, under specific corresponding circumstances. The real dilemma is not "To be *or* not to be," but "To be *and* not to be."

Karl Popper (1968a) emphasises the importance of the method of trial and error in the logic of scientific discovery and looks at Hegel's dialectic in a less religious or mystical way. He explains the dialectic triad, i.e., thesis, antithesis, and synthesis, in a rather profane, as a matter-of-fact way: "First there is some idea, or theory, or movement which may be called a 'thesis'. Such a thesis will often produce opposition, because, like most things in this world, it will probably be of limited value and will have its weak spots. The opposing idea or movement is called 'antithesis'. The struggle between the thesis and the antithesis goes on until some solution is reached which, in a certain sense, goes beyond both thesis and antithesis by recognizing their respective values and by trying to preserve the merits and to avoid the limitations of both. This solution is called the 'synthesis'. Once attained, the synthesis in its turn may become the first step of a new dialectic triad. The dialectic triad describes fairly well certain steps in the history of thought... There are [however] many instances of futile struggles in the history of human thoughts, struggles which ended in nothing." The dialectic logic asserts that contradictions cannot be avoided since they occur everywhere in the world, invalidating the traditional classic logic according to which two contradictory statements can never be true together, or that a statement consisting of the conjunction of two contradictory statements must always be rejected as being false. As Karl Popper remarks: "Dialecticians are right only so long as we are determined not to put up with contradictions and not to change any theory which involves contradictions... Pointing out of contradictions induces us to change our theories, and therefore to progress." The point of view of this monograph is that there is a contradiction between the dialectic logic and the classic logic only if the time and the context are ignored. The dialectic logic is right in emphasizing the fact that there are contradictions everywhere and the classic logic is also right that two contradictory statements cannot be both true if, in both cases, the time and the context are ignored or are both frozen and kept fixed. Pointing out contradictions is fertile because, trying to resolve them, i.e., analyzing the circumstances, the context, and the conditions under which the contradictory statements are true, possible, or false, with certainty or with some probability, we improve our existing theories and we may progress in our understanding of a part of this world. Therefore, the contradiction between the dialectic logic and the classic logic may be solved by using the relative logic. In such a case, the dialectic triad would be: the classic logic as "thesis", the dialectic logic as "antithesis", and the relative logic, essentially taking time, context, and probability into account, as "synthesis".

1.3. Surprise

One day, I decided to sit down and write an exhaustive essay about surprise. With each written page, it became more and more obvious that such an essay could not be exhaustive at all. Surprise is so complex and so deeply infiltrated in our way of living, thinking, or experiencing that deciphering all its features would be an illusion. Therefore, instead of a comprehensive study about surprise, here are only some pages of reflections on ways in which surprise is affecting us or is created by us in positive or negative ways. Surprise appears when our conclusions based on classic logic are contradicted by new circumstances and an unexpected, new context. Our fascination with surprising facts is unbelievable. The first news in newspapers, on radio, or on television is related either to very rare or highly improbable events. If an airplane is highjacked or crashes somewhere, the news is amply discussed for a couple of days, whereas the fact that on the same day ten thousands carriers have safely landed is too banal to be mentioned. Perhaps this mainly reveals our concern for stability and predictability, so necessary for our survival; perhaps it has also something to do with our desire to learn from anything new and unexpected. Almost everybody knows how little chances really are for winning the lottery and still so many people buy a ticket every week. Almost everybody is worried about wars or crimes and still there is a large audience watching the daily movies involving murder and violence. We travel mainly because we want to see something unseen before and we like to listen to or narrate about surprising and unusual things, people, or facts.

Webster's New Collegiate Dictionary tells us that to surprise, astonish, astound, amaze, shock, or flabbergast, all mean to impress forcibly through unexpectedness, startlingness, or unusualness. Surprise tends to enhance the intensity of the emotions. It is suggested that surprise tends to occur when an expectation, conscious or nonconscious, is frustrated or when the stimulus causing surprise conflicts with, or is incompatible with, previous knowledge, belief, or experience. This is a scholarly way of talking. Much easier would be to start from a simple story.

Prediction paradox.

The prediction paradox, or surprise examination problem, is the following one: A teacher announces to her class that there will be an examination on exactly one of the five work days of the next week and that the examination will take the students by surprise – the students will not be able to predict with certainty, prior to the day of the examination, on which day it will be held. The students claim that this announcement cannot be fulfilled for if the examination is left until the last day, Friday, the students will be able to anticipate it on Thursday evening: if that day is ruled out, the same reasoning will apply to the next earlier day, and so on until all the days are eliminated. So, according to students' analysis, a surprise examination is impossible. But in spite of this, the announcement is fulfilled when the teacher gives the examination on one of the days, thus catching the students by surprise.

The flawless reasoning of the students is somehow rudely brought to nothing by the actual occurrence of the promised, but apparently impossible examination; "It has," as somebody said, "the flavour of logic refuted by the world." There were numerous attempts at solving this paradox. Some tried to find errors in students' slippery argument. Others introduced some additional, lateral suppositions on how the students ought to have thought correctly for realizing that a surprise examination was nevertheless possible. Somebody could say, for instance, that the students cannot know whether the teacher will keep her promise or not and that, consequently, they will not know ahead of time whether they are to have the examination or their teacher is to be proved a liar. This looks like cutting the Gordian knot. Making suppositions like that, no surprise examination will be possible if, for instance, the teacher becomes ill or if the school closes unexpectedly because of a severe snowstorm. In what follows we are dealing with the classic formulation of the prediction paradox, as presented above. No additional supposition will be allowed. Under the given constraints, is a surprise examination possible? Can the teacher who promises to give a surprise examination during the following week keep her word? Is students' backward induction used for rejecting the possibility of a surprise examination correct or not?

In spite of numerous attempts, the general opinion is that no simple irrefutable solution to this paradox has yet appeared. We leave its solution to the reader but not before taking advantage of the prediction paradox for distinguishing between different kinds of surprise possible.

Suppose that Mr. Brown is waiting at a bus stop and he knows that the next bus may be anyone from the set of buses labeled 41, 60B, 117, or GARAGE, passing by. He cannot predict with certainty what bus is going to come first. The arrival of one of these buses is a *type I*, or *mild, surprise*. Based on past experience, Mr. Brown could even roughly know how frequent these four types of bus are. In such a case, the arrival of a very frequent bus will be a small surprise for Mr. Brown, whereas the arrival of a rare bus, a big one. If one day, at the same stop, the arriving bus proves to be labeled 57 and Mr. Brown is not aware of such a possibility, he will experience a *type II*, or *shocking, surprise*, totally unpredictable, of course. Including such a new outcome in the set of possible outcomes, he will continue to experience only type *I*, or mild, surprises on the following days until something totally unpredictable will happen again. Therefore, the occurrence of an outcome priorly known as being possible produces a type *I*, or mild, surprise; the smaller the frequency of the possible outcome, the bigger the surprise produced by its occurrence. Moreover, the occurrence of new, priorly unknown as possible, outcomes produces a type *II*, or shocking, surprise.

Before dealing with the prediction paradox as formulated above, let us examine briefly two simple variants of it. The constraints will be chosen from the following ones:

(A) Exactly one examination will be given sometime during the next week.

(B) The examination will be a surprise, i.e., the students will not be able to predict with certainty, prior to the day of examination, on which day it will be held.

(C) At most one examination will be given next week.

Of course, (C) is a weaker restriction than (A). We assume that: a) The teacher informs the students about the constraints chosen by her and acts according to the selected constraints: b) The students are informed about the selected constraints and are confident that the teacher will act accordingly. Otherwise, a surprise may arise either from teacher's inconsistency or from students' suspicion. Also, we are interested here in the existence of an objective surprise examination. Subjectively, the problem is much more delicate. Thus, a good student, prudent and always well prepared for a tomorrow examination, will have, subjectively speaking, no surprise at all from an examination on any day of the week.

In what follows, we call surprise examination an examination that cannot be predicted with certainty. Let us denote by $\mathcal{E}$ the following experiment: "choose a day for examination."

Variant 1: Assume only constraint (A). Abbreviating the work days of the week by *M, Tu, W, Th, F,* and denoting "examination" by 1 and "no examination" by 0, the possible "histories" allowed by constraint (A) are the five rows of the following table:

#	M	Tu	W	Th	F
1	1	0	0	0	0
2	0	1	0	0	0
3	0	0	1	0	0
4	0	0	0	1	0
5	0	0	0	0	1

Obviously, in such a case any work day of the week is a possible outcome of experiment $\mathcal{E}$. In the evening of each day, the remaining possible histories are equally likely. On Sunday evening experiment $\mathcal{E}$ has five possible outcomes, namely the five work days of the week or, equivalently, the five possible "histories" listed in the above table. If the examination does not occur on Monday, then, in the evening of the same day, experiment $\mathcal{E}$ will have four possible outcomes, namely the histories 2,3,4,5 from the above table, and so on. Finally, if the examination does not occur on any of the first four work days of the week, then on Thursday evening experiment $\mathcal{E}$ will have only one outcome, history #5, and there will be no uncertainty about the inevitable examination on Friday. In this variant, the examination will be a surprise examination on any work day of the week but Friday.

Variant 2: Assume only constraint (C). In such a case, there are six possible "histories", shown in the following table:

#	M	Tu	W	Th	F
1	1	0	0	0	0
2	0	1	0	0	0
3	0	0	1	0	0
4	0	0	0	1	0
5	0	0	0	0	1
6	0	0	0	0	0

Again, any work day of the week is a possible outcome of experiment $\mathcal{E}$. The difference is that in Variant 2 the examination will be a surprise examination on any work day of the week, including Friday. If the examination does not occur on any of the first four work days of the week, then on Thursday evening there still will be an uncertainty on a possible examination on Friday.

Therefore, in both variants just presented, a surprise examination is possible, without explicitly asking for it! Notice also that a type I surprise is involved in both variants. But let us go back to the prediction paradox in its initial formulation and let $\mathcal{E}$ be again the experiment: "Choose a day for examination." This time, *the constraints are* (A) *and* (B). The "histories" allowed by constraint (A) are the rows of the first table given above. As we have seen in Variant 1, if only constraint (A) is imposed, then any work day of the week is an outcome of $\mathcal{E}$. But, according to constraint (B), an outcome of $\mathcal{E}$ must be a day on which a surprise examination (not simply an examination!) could occur. Therefore, Friday is not a possible outcome because, according to (A), on Thursday evening there is no uncertainty on what will happen on the next day. Expelling Friday from the set of possible outcomes of experiment $\mathcal{E}$ subject to both (A) and (B), we see that if nothing happens on the first three work days of the week, then on Wednesday evening we will know that an examination can occur on Thursday but not a surprise examination. Indeed, if nothing happens on the first three work days of the week, then the last possibility for a surprise examination will be Thursday and, according to (A), such a surprise examination must necessarily occur on Thursday; therefore the examination is no longer a surprise and Thursday must be ruled out as a possible outcome of experiment $\mathcal{E}$ subject to the constraints (A) and (B). Continuing the same analysis for the other days, we see that no work day of the week could be a possible outcome of experiment $\mathcal{E}$ under the constraints (A) and (B). This is what the students correctly do, rejecting the possibility of a surprise examination. But, in fact, the above arguments simply show that the constraints (A) and (B) do not allow a type I surprise examination. At this stage, we know that there is no type I surprise examination on the work days of the next week. In the evening of each day preceding a work day of the week, as a consequence of the correct logical argument presented above, the students are facing a strictly deterministic experiment with one certain outcome, namely, "No surprise examination tomorrow," which more accurately should be read: "No type I surprise examination tomorrow." But the teacher comes, on any day, and does give an examination, creating a type II surprise. If, for instance, the teacher decides to give an examination on Wednesday, she generates a new outcome, namely, "Examination on Wednesday," to students' pattern which contains only one (certain) outcome "No examination on Wednesday," therefore giving a surprise examination.

Consequently, making a clear distinction between a type I surprise and type II surprise, we see that the students are right in proving that the constraints (A) and (B) of the prediction paradox do not allow a type I surprise examination. But they stop their analysis at this point, unable to predict the possibility of a type II surprise examination which is going to be given by the teacher. Now it is clear why the surprise examination is called a paradox. Both the students, rejecting the possibility of a type I surprise examination, and the teacher, giving a type II surprise examination, are right. But a type II surprise examination is a surprise anyway! Before ending this section, let us mention that the constraints (B) and (C) imply the existence of a type I surprise examination on any work day of the week. Indeed, if constraint (B) is added to (C), Friday continues to be a possible outcome of experiment $\mathcal{E}$ and nothing changes in Variant 2 analyzed above. If the teacher does not use the adjective "surprise" in her announcement and simply says "There will be an examination sometime next week," then she will weaken her commitment to surprise her students. With such an announcement, if she gives the examination on a randomly chosen day, she can garantee that the examination will be given, but there is a probability of $1/5$ that it will not be surprising when given on the last day of the week. Such a case is not paradoxical. Neither is the case when the announcement is: "Perhaps there will be an examination sometime next week." This time the final day of the week cannot be ruled out and the backward induction for eliminating the possibility of a surprise examination cannot be used. The paradox comes up when the announcement refers to giving with certainty a surprise examination sometime next week. The students correctly rule out the last day and, going backwards, all the other days of the week. But the examination will however surprise the students on any day of the next week, including the last day, because they will have eliminated it as a possible event and will no longer expect it.

1.4. Humour

Humour is an excellent illustration of what happens when a rigid classic logic is used ignoring the different contexts and viewpoints possible. Considerable importance has been attached to surprise in the genesis of laughter. Surprise is deemed to be involved in experiences connected with wit, jokes, humour, comic, ridicule, irony, and sarcasm. Apparently amusement is one of those emotions which are increased in their intensity by surprise. Obviously, surprise is not always associated with laughter or amusement. If surprise is followed by curiosity, disappointment, or puzzlement, then surprise may not be associated with laughter or amusement. Surprise leads to a state in which for the moment nothing can be done and laughter results as a relaxation when the subject exposed to the surprising stimulus feels that there is no need to make any further adjustment to the present condition. Aristotle attached considerable importance to surprise in rhetoric and in connection with rhyme and poetry as he did in the case of wit. Apparently Arthur Schopenhauer, the German pessimist philosopher, said that humour is the turnover of expectation. Since the surprising event becomes focal in consciousness, demanding a convergence of mental energy towards itself, it can be seen why it tends to free the mind of what before occupied it. At the beginning of the 20th century, A.F. Shand regarded surprise as a "frequent cause of forgetfulness", as a way of accommodating ourselves to the shock produced by the surprising stimulus that breaks up our expectations.

The dictionaries describe humour as being a sudden, unpredictable, or unreasoning inclination. It may have a good or poor taste. Having a good sense of humour is a quality. A normal question followed by an unexpected, out-of-context answer, incongruous with the content, make us relax and smile. The majority of the jokes from this chapter are certainly not new. They have been heard or read at different times from at least two different sources for each of them. The reader is invited here to think about what makes them funny. More intelligent jokes may be found in Paulos (1980, 1985), and Devine and Cohen (1992).

    –"What do you do for a living?"
    –"I live with my aunts."

    –"How did you spend $200,000 loan from the bank?"
    –"Quickly."

    –"When do you collect the crop on your fields?"
    –"Always in the afternoon."

    –"Where do you live?"
    –"At home."

–"Are dead people buried with a priest in your village?"
–"No, in our village the priest remains outside."

The lord: –"George, could you wake me up tomorrow morning at 6 o'clock?"
The butler: –"Of course, mylord, if you will be so kind to wake me up at 5:30."

–"Are you single or married?"
–"Single married."

–"Is the saxophone a brass or a woodwind instrument?"
–"Yes."

The patient: –"Every morning, when I drink my coffee I feel an acute pain either in the right eye or in the left one, but never in both. Could you find a remedy to it?"
The doctor: –"Did you try to remove the teaspoon from your cup before drinking your coffee?"

The employer: –"You seem to be qualified for the job but, nevertheless, you are too young."
The applicant: –"I promise to remedy this on an everyday basis."

–"Do you know what will happen if you don't tell the truth in this court?"
–"Yes, sir. Our side will win."

–"What do you think about Toulouse-Lautrec?"
–"Toulouse will win."

The patient: –"I think I'm an umbrella."
The psychiatrist: –"A cure is possible if you'll open up."
The patient: –"Why? Is it raining?"

–"If you were my husband, I would put poison in your coffee."

–"If I were your husband, I would drink the coffee."

–"Doctor, may I finally take a shower?"
–"Of course you may, but why are you asking me that?"
–"Because six months ago you advised me to avoid humidity."

A man, notorious for his life full of excesses is interviewed by a reporter on the 90th anniversary of his birthday:
–"What is the secret of reaching such an old age?"
–"I did not smoke, drink, or have love affairs before I was eleven years old."

All is destiny; the rest is fatality.

The judge: –"So, is it true that in the last two years you have lived in adultery?"
– "No, sir. In the last two years I have lived in Toronto."

A young man goes home along a country road. Not far, in the middle of a vast corn field, he sees an older man frenziedly agitating his hands.
The young man: –"What are you doing?"
The old man: –"I am rowing in the sea of corn."
The young man: –"You stupid, old fool. Because of people like you our village got the bad reputation of being a bunch of idiots. I would like to go there and punch you in the nose but, unfortunately, I do not know how to swim."

Two hunters are out in the woods when one of them suddenly collapses. He does not seem to be breathing and his eyes are rolled back in his head. The other hunter takes out his mobile telephone and calls the emergency number.
–"My friend seems to be dead. What can I do?"
–"Keep calm. First, let's make sure that he is dead."
There is silence, then a shot is heard. The hunter comes back on the line and says:
–"OK, now what?"

–"Why is television called a medium?"

–"It is neither raw nor well-done."

An optimist believes that we live in the best of all possible worlds. A pesimist fears that this is true.

During a funeral, the best friend of the deceased comes in front of the mourners in order to deliver the eulogy. Before starting, he takes out from his pocket a big, old-fashioned watch, which once belonged to his grandfather, and puts it in front of him on the lectern. The speech goes on and on for an unusually long period of time. When he finally ends the speech, a friend comes and says:

–"The speech was nice but too long. By the way, why did you put your big watch in front of you on the lectern before starting your speech?"

–"Well, you see, I never look at my watch."

Hercule Poirot and his close friend, Captain Hastings, go camping. After a good meal and a bottle of wine they lay down for the night and went to sleep. Some hours later, Poirot awoke his faithful friend:

–"Hastings, look up at the sky and tell me what you see."

–"I see millions and millions of stars."

–"What does that tell you?" asked Poirot.

–"Astronomically, it tells me that there are millions of galaxies and maybe billions of planets. Meteorologically, I suspect that we will have a beautiful day tomorrow. What does it tell you?"

Poirot was silent for a minute, then spoke:

–"Hastings, you do not use properly your grey cells. Somebody has stolen our tent."

In a totalitarian regime, surprise could be generated when somebody (a writer, or simply a joke teller) dares to say publicly something known by everybody about a topic officially classified as taboo.

Thus, in a certain country, the dictator imposes a law according to which $4 + 4 = 15$. The students protest in front of the presidential palace shouting "$4 + 4 = 8$, $4 + 4 = 8$!" The revolt is brutally repressed by the army. After many years of fights, the dictator is finally disposed and the new ruler replaces the former law by stating that $4 + 4 = 9$. The students are again on streets shouting "$4 + 4 = 8$, $4 + 4 = 8$!" The ruler comes in front of the students saying: "Are you crazy? Do you want back the old hard times when $4 + 4$ were equal to 15?"

A listener sends a letter to a radio station asking whether is it true that the great poet of revolution committed suicide and if so, what were his last words? The commentator's answer: "Yes, indeed, unfortunately the great poet of our revolution committed suicide and his last words were: 'Do not shoot, comrades!'."

Three political convicts meet in a prison. The first says: "I am here because I was in favour of Comrade G." The second: "I am here because I was against Comrade G." The third: "I am Comrade G."

A psychiatrist shows one of his patients to a friend of his. During more than an hour, the patient answers all sorts of questions in a brilliant way, revealing solid knowledge, logical thinking, and personal charm. When the patient leaves, the friend asks the psychiatrist what is wrong with the patient. "Well, everything is all right with him except that he claims to be Saint Peter, which is impossible because I am Saint Peter."

A simple rearrangement of the same words could result in a surprising meaning. The following joke makes use of this procedure: "In Country A everything is permitted that is not explicitly forbidden; in Country B everything is forbidden that is not explicitly permitted; in Country C everything is permitted that is explicitly forbidden." The words of sentences in everyday language are not commutative and if, however, we interchange some of them, then the effect may be strange. Thus, on the wall of a building in Basle there is the following graffito: "God is dead." Signed: Nietzsche, 1879. Twenty two years later, on the same wall: "Nietzsche is dead." Signed: God, 1901. And another graffito, this time on a wall of the famous La Sorbonne University in Paris: "It is forbidden to forbid." Statistics shows that marriage is 75% effective in preventing suicide; on the other hand, statistics also shows that suicide is 100% effective in preventing marriage. The melancholic French reflection, "Partir, c'est mourir un peu," becames, by partial inversion, "Mourir, c'est partir un peu," whereas, "The road to hell is paved with good intentions," becomes "The road to heaven is paved with bad intentions." "To be is to do" (Sartre); "To do is to be" (Rousseau); "Dobedobedo" (a well-known song).

Innocent statements get an unexpected twist when they are interpreted with respect to an out-of-context viewpoint:

Bodies in motion tend to remain in motion; bodies at rest tend to remain in bed.

How do the mathematicians build a house? The algebraists do it in groups. The statisticians do it with deviation from the standard. The probabilists, probably not.

A professor gave a True/False test and noticed that one of his students flipped a coin before answering each question of the test. After one hour, everyone else had left except for the student who was still flipping the coin.

The professor: –"If you are just flipping a coin for your answer, why does it take you so long to finish?"

The student: –"Excuse me. I have finished indeed, but I am checking my answers."

Surprise may be induced by absurd. Something is absurd when it ignores what is considered to be rational or logical and therefore expected, being ridiculously unreasonable, unsound, or incongruous, therefore unexpected, correct but ignoring the context, the proportions, and circumstances:

The doctor: –"I have some good news and some bad news. The bad news is that you have only one month to live."

The patient: –"That's terrible news! What is the good news?"

The doctor: –"Well, I am getting married next week."

Many are inclined to call absurd anything they do not understand. In its first acceptation, absurd is something that contradicts the rules of logic. An absurd idea incorporates incompatible elements. It cannot be reduced, however, only to some false and useless statements like "a square circle", for instance. In logic and mathematics, Aristotle's *reductio ad absurdum* or Francis Bacon's *probatio per absurdum* is used to prove the truth of a statement indirectly, by showing the falsehood of its opposite. According to Albert Camus, the absurd is generated by the discrepancy between the desire of clarity of human reasoning and the irrationality of the real world, between the nostalgia of unity and the chaotic diversity of things and life itself. Absurd does not mean the end of thinking. Through absurd humour the logical reasoning affords to laugh of itself. The comic "nonsense" breakes an excessively rigid logical corset. Joking or playing with words, we can turn judgements upside down, establishing absurd logical relationships that induce laugh. The contrast is between what we can logically expect to happen and what really does happen. The absurd comic is obtained by giving up the constraints of the logical and rigid thinking:

   – "What day of the week is it right now on the Mars?"
   – "How does Monday smell?"
   – "What did Hercule Poirot think about Agatha Christie?"

A mathematician was invited to give a colloquium talk at the university of a neighbouring town. By electronic mail she accepted the invitation and sent the title of the talk "Convex Sets" to the organizers of the colloquium. On the respective day, she was very surprised at seeing the lecture hall crowded with all sorts of people. She started delivering her talk and gradually the audience dramatically dwindled to only about a dozen who remained in the room to applaud the speaker at the end of her 50-minute talk. Only later did she realize that the title of the talk had been printed in the local newspaper as "Convicts and Sex".

In a psychiatric institution a group of patients have been having a good time. The fellows have been laughing whenever one of them has been calling out a number. The puzzled visitor is told by the doctor that they are telling jokes and the numbers are simply codes for jokes. Then the visitor calls out "37" but nobody laughs.

   – "Is there no joke with this number?" asks the intrigued visitor.
   – "Of course, there is; but you didn't know how to tell the joke."

Based on past events, a probabilist calculated the probability of a bomb being aboard a plane. It proved to be a small number. Then he calculated the probability of having two bombs aboard the same plane which proved to be a much much smaller number. Then he decided to travel always with a bomb in his suitcase.

Before deciding to refuse to look into any project of perpetuum mobile, the French Academy of Sciences published a report showing that the tremendous majority of such projects had been elaborated during the spring.

Education, home or in school, strongly relies on a large collection of positive, tonic, optimistic, constructive, and moralizing well-meant conclusions, rules, and principles, often based on the so-called popular wisdom. They could be useful but, if repeated too often, they certainly become boring. Surprise comes when the expected message is reversed. The numerous Murphy's Laws offer such examples. Here are some of them:

In any field, anything that can go wrong, will. If nothing can go wrong, something will. Everything will go wrong at one time. Left to themselves, things always go from bad to worse. Given the most inappropriate time for something to go wrong, that's when it will occur. Anything that begins well, ends badly. Anything that begins badly, ends worse. In order to get a loan you must first prove that you do not need it. No boss will keep an employee who is right all the time. Everything depends, nothing is always, everything is sometimes. Complex problems have simple easy-to-understand wrong answers. Successful research attracts the bigger grant which makes further research impossible. A meeting is an event at which the minutes are kept and the hours are lost. The law of committee formation: if you leave the room, you are elected. There is no time like the present for postponing what you do not want to do. The probability of meeting someone you know increases when you are with someone you do not want to be seen with. Of two possible events, only the undesired one will occur. Things equal to nothing else are equal to each other. If a lost thing is found, something else will disappear.

An excellent woman mathematician had a difficult time getting a permanent position in a famous university. At one point in time, the chair of the department asked her to report to him about her work on a daily basis. On Monday, she wrote briefly: "Tried to prove theorem." On Tuesday: "Tried to prove theorem." On Wednesday: "Tried to prove theorem." On Thursday: "Tried to prove theorem." On Friday: "Theorem is false."

Mike, the son of a rich banker is a student in the music department of a famous university, doing composition. Having enough money, he has preferred to enjoy life instead of studying music and now he is in a difficult position because on the next day he has to go in front of an exam committee and present there a new composition required in order to graduate. To make things even worse, the president of the exam committee is going to be his declared enemy, the tough professor Charles Gun, whose recent composition entitled "Rivers' Waves" has proved to be a hit. Mike sits in front of the piano in his studio and, just as in the last couple of days, he has no inspiration whatsoever. After several fruitless hours, a striking idea pops into Mike's head. He calls his personal secretary and asks him to sit at his own desk and copy the partition of "Rivers' Waves" in the reverse order, from the end going backwards to its beginning! And satisfied by his brilliant idea, Mike goes to bed to take a well-deserved rest. On the next day, he takes with him the partition written according to his instructions by his secretary and goes to the music department in order to pass the composition exam. When his term comes, Mike bows in front of the examination committee, sits in front of the piano, puts the partition in front of him, and starts playing his new composition. And the stupefied members of the committee hear the melodious sounds of Ludwig van Beethoven's Moonlight Sonata.

Before ending this section, let I give two excerpts from one of Woody Allen's book. The first one (Allen, (1980, p.51)) is about the two-valued logical thinking: "Death is a state of non-being. That which is not, does not exist. Therefore death does not exist. Only truth exists." The second excerpt (Allen, (1980, p.81)) is from his speech to the graduates: "Mankind faces a crossroads. One path leads to despair and utter hopelessness. The other, to total extinction. Let us pray we have the wisdom to choose correctly."

1.5. Mystery

In coping with mystery, a temporal, relative logic is needed, dealing with the probability, ambiguity, plausibility, and belief of different subsets of possible suspects, based on the evidence accumulated at different time moments. Eliminate the impossible, as Sherlock Holmes would say, and whatever remains, however improbable, must be the truth. The mathematical formalism will be given in Chapter 5. Here we simply give some examples of different ways of creating mystery stories by classics of this category. In all these examples the main concern of the authors was to delude the reader and to defy the classic logic.

A series of short movies presented weekly on TV under the title *Alfred Hitchcock Presents* allowed less than 30 minutes to create a surprise; sometimes a brutal one. In one of them, a rude and dangerous woman is sentenced and put in a maximum security tough prison. She wants to escape at any price. The only possibility seems to be offered by Mike, an old employee in charge with burying the corpses of the dead prisoners in a yard behind the prison. He teaches the woman to wait until somebody dies, go to the morgue of the prison, and sneak into the coffin beside the dead; on the next day, after the burial, he will unearth and free her. Everything goes as planned and the coffin with the woman and the corpse is buried in the yard. But nothing else happens and the woman is losing her patience, lit a match and realizes that she is laying beside the corpse of Mike.

In the episode *Fogbound*, of the same series, Karen is alone in a big house. It is late in evening, her husband is away and she is waiting for her boyfriend Simon to come to visit her. Outside is cold and foggy. The friend does not come but, in the middle of the night, a strangely looking man, called Sanchez, rings the bell asking for some gas or at least a telephone to call somebody for help. His girlfriend, Maria, is left outside, waiting in the car. Afraid, Karen does not open the door. On the next day, three young men, friends of Karen, inform her that a girl was found dead on the near beach. Police arrives and investigates. Sanchez claims that during his absence Maria was dragged from the car and killed by somebody. Karen has remorses for not helping the couple. Next evening, Sanchez sneaks in Karen's house wanting to kill her. He is hiding while Simon arrives unexpectedly. He gives no explanation for his absence last night, behaves strangely, and in the end intends to rape Karen. The bell is ringing and three good looking young men arrive worried about Karen's safety. Karen feels relieved and tell them about Simon's attack. The three young men are looking for Simon who has disappeared in the meantime. After pretending to search the house, they disconnect the telephone and it is their turn to attack Karen, confessing at the same time that they murdered Maria. Sanchez shows up, kills the killers and wounds Simon hidden under a curtain. The common surprise generating scheme – the bad is good and the good is bad – is replaced by: the bad (Sanchez) is not so bad, the good (Simon) is bad, and the very good (the three young good looking men) is the real devil.

In fact, in his own films, like *Strangers on a Train*, *Rear Window*, or *Suspicion*, Alfred Hitchcock went much deeper, creating the suspense from the coexistence of good and evil in every human being. Therefore, not bad *or* good, but bad *and* good, in different proportions. The surprise of questioning the purity of our own thoughts and desires. He does not exempt himself from such an ambiguous self- characterization. In one of his many short appearences on the screen, here is Hitchcock, the fat little bonhomme, apparently unable to hurt a fly, deadly shooting his duel adversary in the back and saying: "This is not a genuine duel; the bullets are real, the man is dead, but the doctor is a fake; he is not a doctor but an actor. Better be safe than dead."

In another episode of the series, the characters are: H = the host; W = the wife of H; F_1, F_2, F_3, F_4 = the four friends; C = the cop; B = the butler. The action: F_1, F_2, F_3, F_4 are invited to a party organized by H and W. Wine is served in six identical glases, all taste the wine, and H collapses dead. C shows up explaining that he had been hired by H to watch on the guests. He starts investigating the murder. In their testimonies, all four friends prove to have had a great dislike for H. W confesses that she had been unfaithful and made love with all four friends of H. The friends leave the room and suddenly H stands up proud of his joke. W seems to be terrified by her useless confession. H celebrates his success in finding out the truth about his wife. Suddenly he collapses again and W embraces C who appears to be her partner in the plot to kill H. When she is extatic looking at the dead husband, H rises again from the simulated death and C, who proves to be indeed in the service of H, tides up W and is prepared to call the police for arresting her. H and C take a bottle of red wine hidden in a cupboard and drink, celebrating the end of the show. Both become instantly sick due to the strong poison from the wine. This is no longer a joke. H and C are dying for real but not before seeing B showing up and embracing W for celebrating their final triumph.

In another episode of the series, the characters are: S = a student; D = the dean of the medical school; G = the girlfriend and colleague of S; M

= an employee working at the morgue of the medical school. The action: S, an arrogant young fellow, has to graduate and become a physician as the condition imposed by his ambitious father for inheriting one million dollars. Only one final exam is left to be passed but, unfortunately, just with D who, well aware of S's incompetence and indolence, wants him expelled from the medical school. S tells his girlfriend G about his problem and she offers to help him by tutoring him during the preparation of the final exam. But S has other plans. He meets across M in a pub on the campus. M tells S that he failed to graduate the medical school, long ago, due to the intransigence of the same D. S tells M about his intention to kill D, asking M to help him get rid of the corpse. M agrees, provided that S gives him his expensive red car as a reward. S kills D and, together with M, they throw the corpse in a lake. S closes the deal giving the keys of his car to M and meets G for celebrating the end of his worries. The next morning police are on campus and S is arrested. In his red car they have found the corpse of D and the bloody brik used by S to kill D with his fingerprints on it. The movie ends by showing how S is taken to prison while M, the new dean of the medical school, drinks champagne with G.

In a short movie, *The Defendant*, from the series starring Andy Griffith playing the brilliant and gourmand lawyer Ben Matlock, a businessman, named Gordon, is accused of murdering his associate Al. Gordon, a charmer and womanizer, convinces Matlock's daughter Lyanne McIntyre, a lawer herself, to take his case and succeeds in making her believe in his innocence. The only piece of evidence against Gordon is provided by his secretary Jane, who declares that she had seen him around the place of the killing approximately at the time when the murder was committed. Lyanne, however, proves to the jury that Jane and the victim's brother have been lovers. She lied in her testimony in order to cover the true murderer, her lover, who would have inherited nine million dollars if Gordon had been found guilty. Gordon is found not guilty by the jury and leaves for Florida with Lyanne to savor their success. During their romantic encounter, however, Lyanne discovers traces of Al's blood on the Gordon's nightgown and realizes, too late, that Gordon did kill Al. She tries to reopen the case but without success. Gordon has destroyed the evidence found by Lyanne. The movie ends with Gordon and Jane in bed savoring their success and planning how to spend nine million dollars. But would they trust each other?

In a well-done movie, an international group of very different and picturesque people laboriously plan to steal a famous and precious dagger from the Topkapi museum in Istanbul. As the dagger is kept in the main room of the museum under a glass bell surrounded by a sophisticated alarm system, the last phase of the action, after many other breath taking moments, involves a slim acrobat hanging from the ceiling, lifting the glass bell, and removing the shining dagger without touching the dangerous, alarm filled floor. But just when the ultimate objective is so close at hand, all the long and careful preparations made by the group for preventing any possible surprise are ruined by the unexpected appearence of an innocent dove who enters the room through the open little window left ajar by the acrobat and lands on the floor searching for some seeds to eat.

The pattern mentioned above has been used many times. In an old suspense comedy movie, *The League of Gentlemen*, an ex-British army officer recruits other ex-officers to do a bank caper with military precision. They try to foresee and face all possible events, carefully observing and taking photographs of the daily arrival of the armored car loaded with money to the bank targeted by them. The lack of surprise in the daily routine used by security and bank officers helps the team in their detailed planning of the heist. Everything goes well but, helas, a new, unexpected event occurs: an innocent little boy writes down the number of the plate of the car of the leader of the hold-up group. Police arrest the group to the disappointment of the audience who is prevented from enjoying a surprise end, favourable to the nice robbers. By the way, there is in this movie a remark about surprise. In the night preceding the great day planned for the attack on the bank, the members of the group of robbers cannot sleep, like in the old times on the eve of decisive battles, and one of them says: "In love we always expect to be different. And it is always the same. It's the sameness that surprises."

Normally, in a movie where the main character is presented during a long time period of his/her life, several actors of different ages are cast to play that character. Surprisingly, François Truffaut chose the opposite strategy: he took one actor, Jean-Pierre Léaud, and made five movies with him, waiting for the actor to grow up from childhood (*Les quatre cents coups*) to adolescence and maturity (*Antoine et Colette, Baisers volés, Domicile con-*

jugal, and *L'amour en fuite*).

In the excellent novel *The Pledge*, written by Friedrich Dürrenmatt (1958), a little girl is murdered in a Swiss village. To make the story short, after many events, analyzing several similar murders that had happened in the region, Constable Matthäi arrives to the right conclusion that the murderer had to be a psychopath who felt the need to kill little girls periodically. Having confidence in his hypothesis, he prepared a trap, letting a little girl wondering around apparently unsupervised, in order to attract the maniac. But nothing happens because in the meantime the muderer died. The new attack does not take place and the reputation of Constable Matthäi is ruined. A correct hypothesis, made after detecting a pattern in past similar events, is not confirmed because suddenly the context has changed.

It is well-known that generally a mystery novel is created by its author starting with the end of it. The characters and the details of the plot are conceived first. Afterwards, the story is written from its beginning towards the end with the essential care of hiding the truth as long as possible beneath a proliferation of irrelevancies, deceiving the reader and creating a chain of surprises culminating with the final discovery of the real culprit. How uncertainty is created for the reader in order to conceal the truth is a matter of talent. The more uncertainty the greater the final surprise at finding the truth. The reader is engaged in a kind of intellectual puzzle-game and very often wants to discover the solution of the mystery before reaching the final pages of the story. The reader, however, could guess what has happened only if the culprit belongs to the set of characters presented in the novel. Therefore, in order to give the reader a chance, the author of a mystery story is allowed by the unwritten laws of fairness to create only a mild surprise. A strong surprise, i.e., the murderer is not one of the characters involved in the story and is brought up by the author in the last minute out of his pocket, would be unpredictable and therefore unacceptable to the reader. Thus, the author creates a mild surprise to the reader and the murderer is one of the characters of the story. Even so, there are many possibilities open for deceiving the reader. A mystery story is good if the author succeeds in creating a lot of uncertainty in the process of identifying the criminal. Often the culprit is the least likely suspect but sometimes he/she is just the most likely suspect, in which case the author's strategy is to make the reader refuse to accept the obvious. The main interest of the detective story is not crime itself but the tortuous process of solving the problem and discovering the truth by looking for evidence and using the "little grey cells", as Hercule Poirot would say, in order to eliminate the uncertainty about which character is the real culprit.

Agatha Christie (1890-1976), "the queen of crime" and "the mistress of fair deceit", used an impressive variety of ways for creating uncertainty in her detective stories. From 1915 to 1973 she wrote 67 novels and 117 short stories. Hercule Poirot, the little Belgian detective, eternally twirling his moustache and tilting his egg-shaped head, is present in 33 novels and 52 short stories; the clever old Miss Jane Marple, the village spinster who sees all and knows all, in 12 and 20, respectively. Agatha Christie could complete a detective novel in about six weeks, working directly at an old-fashioned typewriter during most of her career. She did not invent the "least-likely suspect" technique but skilfully refined it to become in fact "the one never suspected" variant of it. Among her greatest talents is her ability to vary her murderers so as to conceal them from the reader among the other characters; anyone may be the murderer in her work. From her impressively rich mystery production we are going to focus briefly on only three novels. Agatha Christie plays fairly and creates only a mild surprise. The culprit does belong to the list of characters but the author shows the most ingenious ways of creating an uncertainty for the reader on the true identity of the culprit. This quality of unexpectedness makes Agatha Christie unique among crime writers. Even a relatively modest novel, whose title is the last statement made by a dying man at the beginning of the story, namely, *Why Didn't They Ask Evans?*, keeps us uncertain about who the mysterious Evans is until the last pages of the book. After reading several mystery novels written by her, the conclusion is obvious: "It is the reader's business to suspect absolutely everybody."

From Agatha Christie's rich mystery production we are going to focus on only three novels, namely, *The Murder of Roger Ackroyd*, *Murder on the Orient Express*, and *And Then There Were None*. It is well-known that a critic of detective fiction is under the unwritten obligation not to reveal the resolution of the story. As we are here mainly interested in discussing Agatha Christie's ways of creating a surprise for the reader, it is impossible not to reveal her solutions. Thus, if you did not read the three novels mentioned above, you are strongly adviced to skip the rest of this section until you finish reading the respective novels up to their ends. More details about Agatha Christie's novels may be found in Barnard (1980), Maida and Spornick (1982), and Osborne (1982). .

1. *The Murder of Roger Ackroyd* (written in 1926). It is considered to be the most ingenious crime novel written by Agatha Christie. She herself usually mentioned it as among her three or four favourites. The story is narrated by the local doctor Sheppard. It begins with the death of someone other than Roger Ackroyd. Mrs. Ferrars, a wealthy widow, has been found dead in her bed and Dr. Sheppard has been sent for. He suspects suicide, but sees no point in saying so publicly. The following evening, Roger Ackroyd, a wealthy widower whom village gossip had prophesied would marry Mrs. Ferrars, is murdered in the study of his house. Dr. Sheppard's sister, Caroline, an acidulated spinster full of curiosity, knowing everything and hearing everything, keeps house for him. Their neighbour is just Hercule Poirot, the retired detective growing vegetable marrows. When Poirot is asked to investigate the murder of Roger Ackroyd, he allows Dr. Sheppard to be his assistant and part-confidant. The suspects are many. Most of them were staying in Ackroyd's house when he was murdered. They are: Major Blunt, a big-game hunter, is an old friend, and appears to have a romantic interest in Ackroyd's niece Flora; Flora and her mother, who is Ackroyd's widowed sister-in-law, are poor relations living on Ackroyd's charity; Geoffrey Raymond, the dead man's secretary; Ursula Bourne, a palourmaid; Ralph Paton, Ackroyd's adopted son who is burdened with gambling debts. As in almost all her novels, the characters are convincingly presented and all come under suspicion. Apparently, Agatha Christie wrote a story so perfectly typical of the period. A country squire is murdered, the body is found in the library, there is a butler and a housekeeper both of whom behave suspiciously, the cast of characters includes several with a motive for murder. This typical story is the trap the reader falls into by traditionally assuming that the murderer seemed to fit so naturally into such a milieu. The reader eliminates the real murderer from the list of suspects and therefore gives to Agatha Christie the possibility of creating indirectly a strong surprise at the end of the novel. For the first time we have the murderer as "aide" to the detective in his investigation. He is both the narrator and a credible character. The narrator's account is true in the sense that no false information is presented. He does not lie; instead he omits certain details thus creating a slight ambiguity with

respect to lapses of time. The reader is misdirected, not expecting someone to participate actively in the revelation of himself as a murderer. Such a twist was labeled as being a "dirty trick" by some critics and "a stroke of genius" by other people. In her Autobiography, Agatha Christie gives credit to Lord Mountbatten who in a former letter to her expressed the idea that the story should be narrated in the first person by someone who at the end of the book is revealed to be the murderer. Apparently, her brother-in-law also said that he would like to see a Dr. Watson (allusion to Sherlock Holmes' well-known assistant and friend) who turned out to be a criminal.

2. *Murder on the Orient Express*, also known as *Murder in the Calais Coach* (written in 1934). The victim is an American businessman named Ratchett. The murder is committed in one of the sleeping compartments in the Istanbul-Calais coach of the Orient Express. Hercule Poirot is on the train, returning from Syria. The Orient Express, snowed up in Yugoslavia, provides an ideal set up for a classic exercise in detection involving an international list of characters. Among the potential suspects are: Mrs. Hubbard, an American who bores everyone with tales of her wonderful daughter; the Count and Countess Andrenyi, a Hungarian diplomat and his wife; the Princess Dragomiroff, an exotic Russian accompanied by Hildegarde Schmidt, her German maid; Mary Debenham, a young English governess; Mr. Arbuthnot, a British Colonel on his way home from India; Greta Ohlsson, a Swedish missionary; Ratchett's secretary; an American, so called private investigator; Antonio Foscarelli, an impulsive Italian; the train conductor. At least some of them are not what, at first sight, they appear to be. Poirot finds out that the killed is a killer. But who is the killer of the killed killer? The convention had grown up that in a detective story it was quite likely that one or several of the possible suspects would turn out to be much more closely connected to the dead person than he or she pretends and eventually is/are proved to be the criminal(s) whereas the other initially suspect characters are proved to be innocent. Agatha Christie comes up with a new variant and makes all the suspects murderers. The complicity among twelve likely suspects produces false clues, none of which seems to point to the identity of the traditional murderer. The possibility of a ritualistic murder, an execution carried out effectively by all twelve suspects, is completely removed from the expectations of the reader. And more than that, in the end they are all let off scot free. Due to the audacity of its solution, this novel is considered by many as

being one of the most outrageous murder mysteries. Agatha Christie relied on the fact that in the name of the golden traditional mystery literature the reader would definitely expect one of the suspects to be the murderer but certainly not all of them.

3. *And Then There Were None* (written in 1939). Probably the best-known Agatha Christie's novel. Seven men (Mr. Justice Wargrave, Captain Philip Lombard, General Macarthur, Dr. Amstrong, a young man Anthony Marston, an ex-detective Mr. Blore, and the butler Mr. Rogers) and three women (the butler's wife Mrs. Rogers, an ex-governess Vera Claythorne, and a sixty-five-year-old spinster Emily Brent) have been invited in written letters to stay for one week in a mansion on a small island by its mysterious owner Mr. Owen. Except the butler and his wife, the ten people are strangers to each other and they do not know Mr. Owen who, by the way, is nowhere to be seen. After the first dinner, a voice in a gramophone hidden behind the drawing room wall charges each guest with the murders of one or several people. Nobody else is on the island and the contact with the outside world is impossible due to bad weather and high waves. In the following days, the guests are killed one by one, according to the lines of the old nursery rhyme displayed in a frame in each guest's room which begins with:

> Ten little boys went out to dine;
> One choked his little self and then there were nine.

and goes on counting down to the last lines:

> One little boy left all alone;
> He went and hanged himself and then there were none.

The deaths of all ten guests are grotesque and well-planned. The police arrive when the storm ends. It is obvious that the ten guests were the only people on the island and no person could have left the island because of the rough sea conditions. Without any exception however, all the bodies appear to be murder victims and not a single one shows evidence of being that of a murderer who eventually committed suicide. Paradoxically, the last little "boy", Vera Claythorne, was certainly a victim and not the murderer. She did not hang herself but was hanged; the chair on which she stood adjusting the rope round her neck was not found kicked over but neatly put back against the wall like all the other chairs. But then, who was the murderer? The plot is strange and unusual. All the characters change position and play three distinct roles: suspect, detective, and victim. It is not a "detective story without a detective", how the novel has been labeled, but a detective story with all its characters looking for the author of the murderous plot. When the first murder is committed, it becomes obvious that each of the remaining nine is in danger. Unable to leave the island or communicate with the outside world, they act together to find the killer. Each is both a participant in the investigation process and, since no stranger has been discovered on the little island, still a suspect until he or she becomes a victim. It is a cat and mouse puzzle-game. The game is complicated by the fact that each of the characters had an obscure past and is capable of committing a murder. One by one the characters are eliminated. When only two people remain, one must be the murderer. However, it does not work out like that. Again, it is the reader who, based on a credulous common sense, eliminates the dead from the list of potential suspects, leaving space for a strong surprise, and Agatha Christie calmly grabs the opportunity. Terminally ill and unable to resist to the inner desire of committing a murder on a grand scale himself, after an entire life spent as a judge, Mr. Justice Wargrave, the sixth "victim", helped by Dr. Armstrong acting as an acomplice, fakes his own death and diverts suspicion. Dr. Amstrong is the next real victim and the reader entirely focuses on the remaining three suspects still alive that continue the cat and mouse chase.

1.6. Guessing, knowing, and learning

Evolution and complexity.

Very often, the way by which a solution is found is the method of trial and error. This method is often used by living organisms in the process of adaptation. The success of such a strategy depends on the number and variety of trials. Generally, the more we try, the more likely it is that one of our attempts will be successful. The process of learning essentially depends on repetition. An old saying states that "Repetition is the mother of learning." Also, after several repetitions, an intelligent individual succeeds in performing better and faster. But why is it so?

The different operations that have to be performed for achieving a certain goal have not the same importance. Some of them prove to be utterly useless and it is a pure waste of time and energy to perform them. Even less bright people succeed in identifying such useless operations, ignoring them afterwards. But even among useful operations with respect to the given goal, some are optimum, i.e., by performing them, the largest amount of uncertainty on the final outcome is removed, and some are not, i.e., by performing them, a certain, but not maximum, amount of uncertainty is removed. An intelligent person, during the learning process, not only eliminates the useless operations, but focuses on the optimum operations as well, saving time and energy in the process of reaching the final goal. The human intelect has this learning capability, achieved after a very long evolution process based on the trial and error strategy.

In pattern-recognition, for instance, let us assume that the final objective is to identify a specific disease out of a set of possible diseases. Using a medical example, these possible diseases are the entities of our pattern-recognition problem. These entities are defined by specifying the logical values of some characteristics, i.e., symptoms and results of some laboratory tests if we refer to the medical analogy, and using the logical operations "and", "or", and "non". Between some values of some characteristics we can have incompatibility relations, which reflect the real fact that some combinations between some values of some characteristics cannot coexist and correspond to none of the known entities. Initially, there is a prior probability distribution on the set of the possible entities (i.e., diseases), which is determined based on past experience. This prior probability distribution reflects the starting uncertainty on the set of possible entities and we want to eliminate it during the pattern-recognition process. The naive approach in pattern-recognition would be to check all the characteristics (i.e., symptoms and laboratory tests) and eventually to identify the respective entity (i.e., disease) for each case (i.e., patient). Unfortunately, this is what happens ordinarily in any hospital or clinic. This takes time, costs money, and is very stressing, physically and psychologically, for the patient. And even when all symptoms and results of the tests are finally available, it is not always trivial for the physician to quickly identify the respective disease. The mathematical analysis shows, however, that not all characteristics are equally important. We can determine which characteristic is the most important, i.e., the verification of what characteristic would remove the largest amount of the prior uncertainty. The pattern-recognition process should start by investigating this optimum characteristic. Depending on the possible value of this optimum characteristic, we select the next characteristic which removes the largest amount of the remaining uncertainty on the set of possible entities. Depending on the possible values of this second best characteristic, we are looking, under the new circumstances, for the next best characteristic, and so on, until the entire uncertainty is removed and the corresponding entity is finally identified.

The entire pattern-recognition strategy is summarized in a tree-graph whose branches exhaust all possibilities allowed by the respective problem. When a particular patient is analyzed, in his/her case we follow only one branch of the tree, but the graph shows us what to do whoever the particular patient is. Such an anlysis does not treat the characteristics as being independent, and normally they are not. The big surprise is that very few characteristics, but not always the same, have to be investigated on every branch of the graph.

Sometimes, at different nodes of the graph, we can have a tie between several optimum characteristics. At such a point each of them is entitled to be selected as the next best characteristic to be analyzed and, depending on our choice, we obtain equivalent pattern-recognition algorithms. The mathematical tool needed for obtaining such a pattern-recognition algorithm consists of the basic rules of the multi-valued logic, showing us what kind of algebra has to be used in dealing with sentences, and of the measures of the amount of absolute and relative uncertainty, chosen for dealing with the essentially probabilistic relationship between the possible entities and the available characteristics. In its ambitious attempt of understanding the human intelligence, the artificial intelligence needs a relative logic.

Chance and necessity are basic features of the universal interdependence. The essence of Charles Darwin's theory is this: populations of living organisms suffer random variations, and successful mutations bestow a selective advantage on the offspring in a ruthlessly competitive world. Over time, better adapted variants gain an edge over less well-adapted competitors. With the twin processes of random variations and natural selection, the enormous variety and complexity of life on Earth has arisen, over billions of years, from some primeval chemical broth. The survival of the fittest. But how to explain the "ladder of progress" in evolution, with microbes at the bottom and humans at the top without referring to God or to the guiding hand of a law of nature? The dilemma is even more troubling if we take into account the fact that all living organisms have the same genetic mechanism for transmitting the genetic information from a generation to the next one. Some, like Jacques Monod (1974), for instance, see each evolutionary step as being a pure accident, "chance caught on the wing". The random mutations are responsible for the diversity of species. Other people, like Lila Gatlin (1972), for instance, believe that trying to increase the redundancy of the genetic message, as a way of coping with random errors, the inferior organisms have done it by increasing the divergence from equiprobability whereas the superior organisms have achieved the same goal in a more subtle way, by increasing the divergence from independence, introducing dependences among the "letters" of the genetic alphabet, similar to the dependences between the letters of the natural languages which, in spite of their tremendous variety, have the same mathematical structure, being characterized by specific transition probabilities between the letters or the sounds of their respective alphabets. In spite of many questions remained unanswered, what is certain is the fact that the logic of life cannot be understood without taking the context, probability, and time into account, the triad which essentially characterizes the relative logic.

The investigation of so-called chaotic systems reveals a deep linkage between apparently random behaviour and the spontaneous appearance of order. There is order in chaos. Pure chaos and perfect order could cause similar dilemmas. The extremes meet. The fact that we are confused by pure chaos and we are not well prepared to cope with it, is common knowledge. Human beings have generally tried to introduce order, connection, and structure in the surrounding environment. Surprizing is, however, that perfect order could have a similar efect as well. A couple of years ago, I attended an international conference held at Lomonosov University in Moscow. The main building of this university is huge, very ramified, and perfectly symmetric. After attending so many and diverse lectures presented in many identical lecture halls at the conference, one of the main concerns was to find cafeteria as quickly as possible. But due to the symmetry of the huge building, this proved to be a very difficult task. After several painful attempts, I realized that the best strategy was to go randomly through the regular but unknown maze of corridors until, sooner or later, the cafeteria was finally found. Apparently, I was not the only one who discovered and used this strategy because, during my random walk towards cafeteria, I met across the same confused participants several and several times in their own random motions towards the same final objective.

Empirical evidence and theoretical hypotheses.

No observation can see the whole and therefore be absolutely right. The theoretical hypotheses are based on new observations but as empirical evidence changes, hypotheses change as well. The evolution of theories on the nature of light offer a good example of dialectic logic and is presented by J. Gribbin (1984) like a captivating detective story. Based on the facts that light rays are observed to travel in straight lines and bounce off a mirror very much like the way a ball bounces off a hard wall, Isaac Newton explained the behaviour of light in terms of particle. This corpuscular theory of light explained very well the geometric optics, namely, light reflection and refraction. Christiaan Huyghens noticed, however, that reflection and refraction may be explained equally well assuming that the light is a wave, giving birth to the wave theory of light. New facts showed that when light passes a sharp edge, it produces a sharply edged shadow. This is exactly the way streams of particles, travelling in straight lines, ought to behave. A wave tends to bend, or diffract, some of the way into the shadow. Three hundred years ago, this evidence clearly favoured the corpuscular theory and the wave theory, although not forgotten, was discarded. Apparently, in the eighteenth century, very few people took the wave theory of light seriously.

In the nineteenth century, however, the status of the two theories had been almost completely reversed, due to crucial new experiments performed by Thomas Young and, a little later, by Augustin Fresnel. Young shone a light upon an obstructing screen in which there were two narrow slits. Behind this obstruction, light from the two slits spread out and interfered, giving alternate bands of light and shade striping on the screen, caused by constructive and destructive interference of waves from each slit. Fresnel dealt with the coloured reflections produced when light shines on a thin film of oil. The process is again caused by interference of waves. Some light is reflected from the top of the film of oil, but some passes through and is reflected back from the bottom surface of the layer. So there are two different reflected beams, which interfere with one another. Because each colour of light corresponds to a different wavelength, and white light is composed of a superposition of all the colours of the rainbow, the reflections of a white light from an oil film will produce a mass of colours as some colour waves interfere destructively and some constructively, depending on just where our eye is in relation to the film. Wave theory became triumphant when James Clerk Maxwell established the existence of waves involving changing electric and magnetic fields. In 1887, Heinrich Hertz succeeded in transmitting and receiving electromagnetic radiation in the form of radio waves, which are similar to light waves but have much longer wavelengths.

At the end of the nineteenth century, investigations of the photoelectric effect by Phillip Lenard and J.J. Thompson showed that electrons can be produced by light shining onto a metal surface in a vacuum. The produced electrons moved faster when a beam of light with a higher frequency was used. In 1905, Albert Einstein applied the equation $E = h\nu$ to the electromagnetic radiation and claimed that light is not a continuous wave but comes in definite groups of particles, or quanta, called now photons. All the light of a given frequency ν, which means of a particular colour, comes in groups that have the same energy E, given by the above equality, where h is a constant. The struggle between the two opposite viewpoints on the nature of light became more and more intense. How could light be made of particles when a lot of experiments showed light to be made up of waves? The struggle between the corpuscular theory (i.e., thesis) and the wave theory (i.e., antithesis) has been solved around 1920 when Niels Bohr formulated the principle of complementarity (i.e., the synthesis) according to which the wave and particle theories of the nature of light are not mutually exclusive to one another but complementary and both are necessary in order to provide a complete description and explanation of the evidence accumulated so far. The connection between these two theories, as unity of opposites, formed the basis of the development of quantum mechanics and the next step after the discovery of the dual nature of light was the discovery of the dual nature of matter (i.e., particle or wave, depending on the context) formulated by Louis de Broglie (1924).

Logic and intuition.

Often logic, understood in a rigid way, and intuition are seen to be two different tools for solving problems and acquiring new knowledge and results. Whereas logic allows us to obtain new results by ingeniously manipulating the existing assumptions, accepted results, and new facts according to the logical rules, intuition is seen as being a more subtle way of making new discoveries by some kind of mysterious, unconscious revelations. Being well inspired appears to be more prestigious than performing a thorough logical analysis. Some people claim that invention occurs by pure chance, sometimes in dreams, whereas the systematic reasoning is based on classic logic.

In a rare insight into the mechanism of invention in mathematics, Jacques Hadamard (1954) acknowledges the brusque and immediate appearance of a solution to a problem at the very moment of sudden awakening. Sometimes a problem is consciously abandoned for a while to be resumed again later, benefiting from the unconscious work of the mind on the problem during the period of time elapsed since dropping the study of the respective problem. Great mathematicians, like Henri Poincaré and Carl Friedrich Gauss, openly mentioned the "sudden flash of light", or the "sudden illumination", experienced by them as a result of the prior unconscious work. This creative inspiration coming very often as a sudden accident, confirms that, apparently, the human brain uses a relative, elastic logic, consciously or unconsciously examining a problem by changing the context or the viewpoint from which it looks at the respective problem and its consequences. As Hadamard puts it: "The unconscious has the important property of being manifold; several and probably many things can occur in it simultaneously. This contrasts with the conscious ego which is unique ... There seems to be a kind of continuity between the full consciousness and more and more hidden levels of the unconscious... The intervention of chance occurs inside the unconscious... Memory belongs to the domain of the unconscious... Subconsciousness [i.e., fringe-consciousness or antechamber of consciousness] denotes superficial unconscious processes, at the frontier between unconscious and conscious." To invent means to choose those combinations which are useful with respect to a goal, explain or predict experimental facts and events, or are simply governed by a sense of scientific beauty and harmony.

Often, human minds are classified in two broad categories: logic and intuitive. In his lecture on the mathematical creation, Henri Poincaré (1913, p.387) wrote that: "[Some sort of mathematicians are] preoccupied by logic; to read their works, one is tempted to believe [that] they have advanced only step by step,... leaving nothing to chance. The other sort are guided by intuition and, at the first stroke, make quick but sometimes precarious conquests." Another great mathematician, Felix Klein is credited by J. Hadamard (1954, p. 107) to have said that intuition, with its mysterious character, is superior to the prosaic way of logic. Unconsciously or not, an intuitive mind shows more freedom and dare in modifying and changing the standard contexts whereas a logic mind is more conservative, trying to get all logical consequences inside a given context rather than changing the context itself. In fact, the separation in logic and intuitive minds is far from being rigid. There is hardly any completely logical discovery and some kind of intuition is necessary at least to initiate the logical work. Even when the logical approach is used, the more spectacular results are obtained when the context is changed. When the context, described by a given set of axioms or postulates, is kept fixed, any new result obtained logically from these axioms and already known theorems could be interesting but is not spectacular. This kind of results could be easier obtained by an automatic device of proving theorems and a computer could do the job better and much faster than we can. When, however, the context itself is changed, then an entire new and unexpected world may be built. When, after two thousands years of universal acceptance, one independent postulate of the Euclidean geometry, according to which through an exterior point we can draw one and only one parallel to a given line, was replaced by its two possible negations, i.e., there is no such parallel or there are infinitely many such parallels, respectively, two different non-Euclidean geometries have been born. All three geometries are valid, depending on the context.

Plausible reasoning.

A mathematician uses a demonstrative logic, based on the classic logic, but the inductive methods used by a physicist, the indirect evidence used by a jurist, the documents used by a historian, and the statistical analysis used by an economist are all special kinds of plausible reasoning. The demonstrative reasoning is certain and definitive. By contrast, the plausible reasoning is uncertain, debatable, and temporary. The demonstrative reasoning and the plausible reasoning, however, do not contradict but rather complement each other. The inductive reasoning is a particular case of plausible reasoning. Many results from number theory, for instance, have been obtained by observation and induction, before being rigorously proved using a demonstrative reasoning. A general hypothetical statement becomes more plausible if it is verified by a new particular case or new evidence.

G. Polya (1954) analyzed the logic of plausible reasoning. To give two represetative examples, the 'modus ponens', 'modus tollens', and the incompatibility relationship used in the demonstrative reasoning, namely:

A implied by B
B true
A true

A implies B
B false
A false

A and B incompatible
B true
A false

become:

A implied by B
B false
A less plausible

A implies B
B true
A more plausible

A and B incompatible
B false
A more plausible

in the plausible reasoning, respectively.

The plausible reasoning may be expressed by using the elementary equalities involving conditional probabilities (Polya, (1954)), namely:

$$P(A)P(B \mid A) = P(B)P(A \mid B),$$

$$P(B) = P(A)P(B \mid A) + (1 - P(A))P(B \mid \bar{A}),$$

$$\frac{P(A)}{P(A \mid B)} = P(A) + (1 - P(A))\frac{P(B \mid \bar{A})}{P(B \mid A)},$$

where $\bar{A}$ is the event non$A$. Consequently, the inequality:

$$P(B \mid A) > P(B \mid \bar{A})$$

implies:

$$P(A) < P(A \mid B),$$

which, interpreting P as being plausibility, show that if a certain circumstance is more plausible together with a certain hypothesis than without it, then the confirmation of this circumstance can only increase the plausibility of the respective hypothesis.

1.7. Games and rational decision

Game theory and decision theory deal with a relative logic in choosing the optimum strategies with respect to a goal under variable circumstances induced by an opponent's actions or by nature's possible states, respectively. Time, context, and probability are essential features of the decision making process. Games of chance have raised the first problems dealing with combinatorial probabilities. Later, game theory has become the first domain of applied mathematics involving human beings, i.e., the players engaged in the game, capable of analysing possible strategies and taking rational decisions, in order to maximize their profit or minimize the loss involved. The interaction between players is an essential feature of game theory. Dealing with players with opposite interests, game theory has formulated the minimax and maximin criteria for selecting the best strategies in the struggle for optimizing the own expected profit. Game theory has been generalized by the modern decision theory, where other criteria are needed when the decision maker faces nature, which is not necessarily a malevolent opponent.

Playing chess against a computer.

In a two-person game each player chooses an action or a series of actions, called pure strategy, from a set of possible pure strategies allowed by the rules of the game. A player, say Player I, experiences a type I surprise when the opponent chooses one of several expected strategies, or a type II surprise, when the opponent comes up with a new, unexpected strategy, either allowed by the rules of the game but never used before or even breaking the rules of the game and forced on Player I.

Obviously, one of the most complex, if not the most complex, two-person game is *chess*. It was thought to have originated from an Indian game in the sixth century but, apparently, some scholars discovered much older chesslike pieces in ancient Egyptian tombs or in china jars found in Uzbekistan, near the Afghanistan border. It is said that the earliest works which mention chess as a type of war game, appeared in India, Persia, and Islam at the beginning of the seventh century. From India, chess arrived in Europe sometime in the middle of the tenth century. It has been played in the present form apparently since the end of the 17th century. Today, the game is still very popular. In Switzerland, for instance, both in parks of big cities like Zürich, Geneva, or Bern, or in small and remote villages like Zermatt, for instance, people play outdoors chess, moving pieces, which are about half a metre tall, while walking on huge chess boards, watched by a large audience of all ages.

The game is played on a board of 8×8 squares alternating black and white. Each player has 16 pieces, the one side white and the other black. In each set, there are a king (abbreviated by K), a queen (Q), two bishops (B), two knights (N), and two castles or rooks (R), defended by a line of eight foot-soldiers or pawns (P).

The king is in check when he is in the danger of being taken by some other adverse piece. If the king cannot be removed out of check, the game is lost. The game is tremendously complex due to an enormous number of possible variants. Each of the pieces can move in a particular way. The king, the most vulnerable piece, can move in any direction, but suffering from arthritis, by only one step at a time, except when castling, which allows him to jump two steps to the right or two steps to the left bringing the rook to its other side simultaneously. A player is allowed to castle only if the king is not in check, neither the king not the relevant rook has moved during the game, no pieces separate the king and the relevant rook, and none of the squares over which the king travels belongs to the range of action of an enemy piece. The standard notation O-O describes king's side castling, also known as the short (or small) castling, whereas castling queen's side is denoted by O-O-O, and is known as being the long (or big) castling. The queen, the most powerful piece by far, can move in any direction, at any distance. The bishop can move at any distance but only in a skewed direction which forces him to act only on squares of the same colour during the entire game. The knight jumps only in L-shape. The rook moves at any distance but only in the straight (north, south, east, west) directions. Finally, a pawn can move only forward, one step at a time, except for his first move which may be a regular one-step move forward or, if he shows initial enthusiasm, a one time two-step move forward. A piece may eliminate only pieces of opposite colour entering his/her unoccupied range of action. A pawn may eliminate an opposite piece and take up its position only when it is located one step to the right or one step to the left of that piece; it stops when the position in front of him is blocked. At the beginning of the game, the chess board is:

8	R	N	B	Q	K	B	N	R
7	P	P	P	P	P	P	P	P
6								
5								
4								
3								
2	P	P	P	P	P	P	P	P
1	R	N	B	Q	K	B	N	R
	a	b	c	d	e	f	g	h

The white set is in rows 1 and 2. The two players choose alternatively one move. White moves first. The standard way of describing a pair of moves made by the two players is:

1. Pd2-d4 Ng8-f6 (at step #1 of the game, Player *I* (white) moves the pawn from square d2 to square d4 and Player *II* (black) moves the knight from square g8 to square f6);

6. Bg5×f6 Pg7×f6 (at step #6 of the game, Player *I* moves the bishop from square g5 to square f6, eliminating (or taking as prisoner) the black piece from square f6, whereas Player *II* moves the pawn from square g7 to square f6, eliminating (or taking as prisoner) the white bishop. The notation Pb7-b8/Q means that the white pawn from square b7 moves to the last square in this column, namely b8, and is transformed into a new white queen. We also denote "check" by + and "defeat" by *mate*.

It is not surprising that chess has been a darling topic for people working in artificial intelligence. World's first computer chess tournament opened in New York on September 1, 1970. In the last twenty years many powerful computer programs have been created. I am not a good chess player but I like the game as an entertainment. I have not studied chess literature and I do not want to spend a long time thinking about the best next move. But, as I have said, I like to play chess from time to time. Some years ago, a former student of mine gave me a diskette containing the software *Chess by Psion*, Version 1.01, for personal computers. It is a marvelous three-dimensional chess program. It may be used at 11 levels of difficulty. I use the level 0 (novice) mainly because I am a novice too and upper levels require more thinking time for computer whereas I do prefer a speedy play. For a couple of months, in spite of choosing always the white, which gives the advantage of the first move, I lost all the games. Almost every strategy chosen by computer proved to be a type *II* surprise for me. I could not cope with it. One day I decided to stick to only one opening, a kind of personal variant of the so called Sicilian defence, and see the reaction of the computer, hoping to detect some patterns in its way of playing which, once classified, would allow me to cope with only a type *I* surprise instead of a type *II* surprise. After learning the possible patterns, the next stage consisted in finding the antidote to each particular pattern followed by the computer in counteracting my Sicilian defence. Obviously, I was a much worse player than the computer but I had the great privilege of being able to learn something about its behaviour and detect its weak moves whereas it could not learn and improve its own way of playing.

It would take too many pages to describe the details of the complete set of patterns chosen by computer as possible responses to my opening strategy. Playing chess only occasionally, it took me a couple of years for listing these possible patterns and learn the weak moves of the computer. Since then, I have been able to win systematically any time I have used the selected opening. After eliminating the possibility of a totally new strategy of the computer, which would have been a type II surprise, by classifying all possible patterns, I had to face only a type I surprise, namely a possible pattern chosen by the computer but used before. Inside the selected pattern, I knew the trivial moves, i.e., the moves that could be predicted with certainty. But every time when I made a certain move by mistake, deviating from my standard variant of the Sicilian defence opening, I experienced a surprise of type II from the part of the computer and I lost!

In general, once a player makes a move, the next move of the opponent is *trivial* if it can be predicted with certainty and *surprising* if it cannot be predicted with certainty. A surprising move at the $(k+1)$th step of the game is of *type I* if it is chosen from a set of possible alternatives already used by the opponent in other plays of the game with the same history up to and including the kth step. There are several degrees of surprise contained by a surprising move of type I, depending on the number of alternatives used before and their frequency distribution. A trivial move has the entropy equal to zero; a surprising move of type I has a positive entropy. The larger the entropy of a surprising move of type I, the more important is the move in classifying the behaviour of the opponent. A surprising move at the $(k+1)$th step of the game is of *type II* if it is a new one, never used by the opponent in other plays of the game with the same history up to and including the kth step. When a surprising move of type II occurs, this means that a new pattern in the behaviour of our opponent has to be taken into account with respect to the history of the game up to and including the kth step.

In what follows, the computer has the black set of pieces. I decided to play the same opening, a variant of the Sicilian defence, in every game. The first three moves of the computer were always the same. After the first play of the game, they have contained no surprise at all, being predictable with certainty. The fourth move of the computer proved to be a surprising move. More often than not, it used **P**e7-e6 followed eventually either by the frequent move **B**c8-d7 (I am calling it the variant A_1) or by the less frequent move **B**c8-b7 (the variant A_2). Rarely, the fourth move of the computer was **B**c8-f5 (the variant B). In variants A_1 and A_2 the computer chose the short castling whereas in variant B, the long one. In each of the three variants, many subsequent moves of the computer proved to be surprising moves either, but the crucial moves in order to find a winning strategy were the ones mentioned above. In each variant, once a winning strategy is found, the computer enters the same traps, learning nothing from the previous plays of the game. It proves to be vulnerable to long attacks on diagonals and has an unjustified appetite for a mutual elimination of the two queens. Towards the end of the game, when the victory of the opponent is innevitable, the computer becomes nervous, deliberately making stupid and useless moves or even sacrificing senselessly its pieces in a suicidal attempt. The computer, however, is an excellent chess player and, even at the "novice" level of difficulty, it will have real chances of winning if the white plays without learning the key moves from the dissection and analysis of many past games played against the computer.

A representative play of variant A_1 is:

01.	Pd2-d4	**P**d7-d5
02.	Pg2-g3	**N**g8-f6
03.	Bf1-g2	**N**b8-c6
04.	Ng1-f3	**P**e7-e6 (a key move of the black)
05.	Bc1-g5	**B**f8-b4+
06.	Pc2-c3	**B**b4-e7
07.	Bg5×f6	**B**e7×f6
08.	O-O	**O-O**
09.	Qd1-d3	**Q**d8-e7
10.	Nb1-d2	**B**c8-d7 (a key move of the black)
11.	Pb2-b4	**P**a7-a6
12.	Pa2-a4	**P**h7-h6
13.	Pe2-e4	**B**f6-g5
14.	Pe4-e5	**B**g5×d2
15.	Nf3×d2	**Q**e7-g5
16.	Ph2-h4	**Q**g5-g4
17.	Kg1-h2	**N**c6-e7
18.	Bg2-f3	**Q**g4-f5

19. Qd3-e3	Qf5-c2 (the black queen enters a deadly trap)
20. Pg3-g4	Ne7-g6
21. Ph4-h5	Ng6-e7
22. Bf3-e2	Ra8-d8
23. Be2-d3	Qc2-b2
24. Rf1-b1	Qb2×b1
25. Bd3×b1	Pf7-f5
26. Qe3-d3	Bd7-c6
27. Kh2-g3	Pf5-f4+
28. Kg3-f3	Rf8-f7
29. Qd3-h7+	Kg8-f8
30. Bb1-g6	Ne7×g6
31. Ph5×g6	Rf7-d7
32. Qh7-h8+	Kf8-e7
33. Qh8×g7+	Ke7-e8
34. Qg7×h6	Pb7-b6
35. Pg6-g7	Ke8-f7
36. Qh6-f6+	Kf7-g8
37. Qf6×e6+	Rd7-f7
38. Qe6×c6	Kg8×g7
39. Ra1-h1	Pa6-a5
40. Pb4-b5	Kg7-g8
41. Rh1-h6	Kg8-g7
42. Qc6-g6+	Kg7-f8
43. Rh6-h8+	Kf8-e7
44. Qg6-g5+	Ke7-d7
45. Qg5×d8+	Kd7-e6
46. Rh8-h6+	Rf7-f6
47. Rh6×f6+	mate.

A frequent play of variant A_2 is:

01. Pd2-d4	Pd7-d5
02. Pg2-g3	Ng8-f6
03. Bf1-g2	Nb8-c6
04. Ng1-f3	Pe7-e6 (a key move of the black)
05. Bc1-g5	Bf8-e7
06. Bg5×f6	Be7×f6

07. Pc2-c3	Bf6-e7
08. Qd1-d3	**O-O**
09. O-O	Qd8-d6
10. Nb1-d2	Pb7-b6
11. Pb2-b4	Bc8-b7 (a key move for the black)
12. Pa2-a4	Ra8-d8
13. Rf1-e1	Pa7-a5
14. Pb4-b5	Nc6-a7
15. Pe2-e4	Pd5×e4
16. Nd2×e4	Qd6-d5
17. Nf3-d2	Rf8-e8 (a bad move and the black queen is lost)
18. Ne4-f6+	Be7×f6
19. Bg2×d5	Rd8×d5
20. Ph2-h4	Rd5-d7
21. Nd2-e4	Pc7-c5
22. Ne4×f6+	Pg7×f6
23. Ra1-d1	Pc5×d4
24. Pc3-c4	Pe6-e5
25. Qd3-f5	Re8-e6
26. Pg3-g4	Rd7-c7
27. Pg4-g5	Rc7×c4
28. Pg5×f6	Pe5-e4
29. Qf5-g5+	Kg8-f8
30. Qg5-g7+	Kf8-e8
31. Qg7-g8+	Ke8-d7
32. Qg8×f7+	Kd7-d6
33. Qf7×b7	Na7-c8
34. Pf6-f7	Re6-g6+
35. Kg1-f1	Rg6-f6
36. Re1×e4	Ph7-h5
37. Rd1×d4+	Kd6-c5
38. Rd4×c4+	Kc5-d6
39. Re4-e8	Nc8-a7
40. Pf7-f8/Q+	Rf6×f8
41. Re8×f8	Na7×b5
42. Qb7×b6+	Kd6-e7

43. Rc4-f4   Nb5-d6
44. Qb6-c7+   Ke7-e6
45. Rf8-f6+   Ke6-e5
46. Qc7×d6+   mate.

The worst version of variant A_2 played by the computer so far is the following one:

01. Pd2-d4   Pd7-d5
02. Pg2-g3   Ng8-f6
03. Bf1-g2   Nb8-c6
04. Ng1-f3   Pe7-e6 (a key move of the black)
05. Bc1-g5   Bf8-b4+
06. Pc2-c3   Bb4-e7
07. Bg5×f6   Be7×f6
08. Qd1-d3   Bf6-e7
09. O-O   **O-O**
10. Nb1-d2   Qd8-d6
11. Pb2-b4   Pb7-b6
12. Pa2-a4   Bc8-b7 (a key move of the black)
13. Rf1-e1   Rf8-d8
14. Pe2-e3   Rd8-d7
15. Bg2-f1   Ra8-d8
16. Bf1-e2   Ph7-h6
17. Be2-d1   Pa7-a5
18. Bd1-c2   Pa5×b4
19. Qd3-h7+   Kg8-f8
20. Qh7-h8+   mate.

Very rarely, the computer used variant B, playing the key move **Bc8-f5** instead of **Pe7-e6**. Apparently a minor change, but in fact the entire winning strategy of the white had to be changed, switching the direction of the main attack from the right hand side to the left hand one. A frequent version of this rare variant B is:

01. Pd2-d4   Pd7-d5
02. Pg2-g3   Ng8-f6
03. Bf1-g2   Nb8-c6

04. Ng1-f3	**B**c8-f5 (a key move of the black)
05. Nf3-h4	**Q**d8-d7
06. Nh4×f5	**Q**d7×f5
07. Pc2-c3	**P**e7-e5
08. Pe2-e3	**O-O-O**
09. Pb2-b4	**B**f8-e7
10. O-O	**N**f6-e4
11. Pa2-a4	**P**h7-h5
12. Ph2-h4	**K**c8-b8
13. Pa4-a5	**P**e5×d4
14. Pe3×d4	**B**e7-d6
15. Bc1-e3	**R**d8-e8
16. Qd1-a4	**P**a7-a6
17. Pb4-b5	**P**a6×b5
18. Qa4×b5	**K**b8-a8
19. Pa5-a6	**N**c6-d8
20. Pa6×b7+	**K**a8-b8
21. Ra1-a8+	mate.

It is obvious that the computer would have become more powerful if the programmer had introduced more randomness (not precision) in its behaviour. This is just one of the main problems in artificial intelligence, namely how to put in computers some degree of fuzziness, allowing more adaptability to an ever changing environment, so specific to a human mind.

Will the computers be better chess players than human beings? Personally, I do not think so. I cannot imagine computers capable of playing the alcoholic chess, for instance, which is a chess variant played on a large board, with glasses or bottles of alcoholic drink replacing the chess pieces. If the opponent's "piece" is captured, the drink it contains must be consumed by the capturer before play proceeds. A great chess master is said to have won such a game by sacrificing his queen in ridiculous fashion at the very outset of the game, the rules of the game forcing the opponent to drink about a quarter litre of cognac contained by the queen.

Pure and random strategies.

A two-person zero-sum game is a model of a confrontation between two intelligent players with opposite interests such that the win of a player equals the loss of the other one. Let $X = \{x_1, \ldots, x_m\}$ be the set of pure strategies of Player I and $Y = \{y_1, \ldots, y_n\}$ the set of pure strategies of Player II. The two players choose their strategies independently and simultaneously. A variant of the game consists of a pair of pure strategies x_i and y_j chosen by the two players, respectively. To each variant of the game $\{x_i, y_j\}$, a number u_{ij} is assigned, representing the corresponding payoff of Player I. The payoff of Player II is $-u_{ij}$. Obviously, u_{ij} may be positive, negative, or zero. The so called payoff matrix of Player I is represented by the following table:

		Player II		
		y_1	$\cdots$	y_n
	x_1	u_{11}	$\cdots$	u_{1n}
Player I	$\vdots$	$\vdots$		$\vdots$
	x_m	u_{m1}	$\cdots$	u_{mn}

The two-person zero-sum games introduced the relative logic based on the maximin and minimax strategies of rational players. Let us give two examples. In the first example the optimum solution is given by a pair of pure strategies, whereas in the second example the optimum solution is expressed in terms of mixed, or random strategies of the two players.

1. *The labour union against the administration.* Assume that the labour union and the administration of a company have been negotiating a new labour contract. After days and days of negotiations, they have come up with the wage increase proposals of \$20/day and \$15/day, respectively. As no part is willing to change its proposal, an arbitrator is brought in. She offers the two opponents two days to think about their final offer knowing that she is going to accept the proposal (rounded to the nearest dollar) of the side that gives the most from its last figure towards reaching a compromise. If they both do not change the last figure or give in the same amount, then she is going to choose the average amount, namely \$17.5/day, as the final figure. What strategy would be optimal for each side of the conflict?

Obviously, each player in this serious game has six pure strategies, corresponding to giving in \$0, \$1, \$2, \$3, \$4, or \$5 from its last figure, going down (the union) and up (the administration). If, for instance, the labour union comes up with the final strategy −$1, which means to ask for a wage increase of $19/day, and the administration chooses the strategy $2, which means to agree with a wage increase of $17/day, then the arbitrator will settle the final figure at $17/day, which means a payoff of −$3 for the labour union, corresponding to a decrease of $3 from the last figure of $20 proposed by it before the arbitration game. Denote the labour union by I and the administration of the company by II. According to the fair rules imposed by the arbitrator, the payoff matrix for the labour union is:

					II			
		$0	$1	$2	$3	$4	$5	*min*
	$0	−$2.5	−$4	−$3	−$2	−$1	$0	−$4
	−$1	−$1	−$2.5	−$3	−$2	−$1	$0	−$3
	−$2	−$2	−$2	−$2.5	−$2	−$1	$0	−$2.5
I	−$3	−$3	−$3	−$3	−$2.5	−$1	$0	−$3
	−$4	−$4	−$4	−$4	−$4	−$2.5	$0	−$4
	−$5	−$5	−$5	−$5	−$5	−$5	−$2.5	−$5
	max	−$1	−$2	−$2.5	−$2	−$1	$0	

Obviously, the union would like to keep its last proposal ($20) unchanged, which would correspond to a payoff equal to $0 in the arbitration game. But the final result depends on the way of acting of both sides. The last column in the above table shows the minimum payoff the administration could impose in the arbitration game corresponding to each possible strategy of the union. Obviously, the union would try to choose its own strategy in order to maximize the minimum payoff the administration can impose on it, by selecting the strategy −$2 which would give the so called *lower value* of the arbitration game equal to −$2.5. On the other side, the administration would like to keep the payoff of the union down. The last row in the above table shows the maximum payoff the union can impose in the arbitration game corresponding to each strategy chosen by the administration. Obviously, the administration would try to choose its own strategy in order to minimize the maximum payoff the union could get, by selecting the strategy $2 which would givew the so called *upper value* of the arbitration game equal to the amount −$2.5. The pure strategy −$2 of the union is called the *maximin strategy*, whereas the pure strategy $2 of the administration is called the *minimax strategy.* A two-person zero-sum game in which the lower value of the game is equal to the upper value of the game, like in our case, has a solution in terms of pure strategies. The common value of the lower and upper values of the game is called the *optimal value* of the game for Player *I*. The pair of optimal (maximin and minimax) strategies of the two players in such a case is called an *equilibrium point* of the game. If a two-person zero-sum game has an equilibrium point and a player chooses the optimal strategy (maximin for Player *I* or minimax for Player *II*), then that player can be sure that the outcome of the game cannot be worse than the optimal value. It could be even better if the opponent does not choose the optimal strategy.

Therefore, in a two-person zero-sum game, if both players choose the optimal strategies, maximin and minimax, respectively, then the outcome of the game will contain no surprise for the two players, being fully predictable. If only one of the players, say Player *I*, chooses the optimal (maximin) strategy, then the actual output of the game could be only a positive surprise for Player *I* in the sense that the actual output of the game will be *at least* equal to the optimal value of the game.

In our example, if the union chooses the maximin strategy −$2 and the administration chooses the minimax strategy $2, then the result of the arbitration game will be -$2.5 for the union and $5 − $2.5 = $2.5 for the administration, which means a final wage increase of $20 − $2.5 = $15 + $2.5 = $17.5/day. If the administration decides to choose the minimax strategy $2 but the union does not choose the maximin strategy −$2 and prefers another strategy, say $0, which means to stick to the last proposal, then the arbitration game ends in a payoff of −$3 for the union and $5 − $3 = $2 for the administration, resulting in a final wage increase of $20 − $3 = $15 + $2 = $17/day. Finally, if both players give up their optimal (maximin and minimax) strategies, then they cannot predict the outcome of the game with certainty.

2. *Two-finger mora game.* Unfortunately, not every two-person zero-sum game has an equilibrium point in terms of pure strategies in order to have only a pleasant surprise to the extent to which the optimal strategy is selected. The two-finger mora game is such an example. The game is very simple. It may be easily generalized to a three-finger or ten-finger game. Two players are involved in the game. They show, simultaneously and independently, one

finger or two fingers. If the numbers match, then Player *I* will get \$1 from Player *II*. If the numbers do not match, then Player *I* will pay Player *II* \$1. How has the game to be played in order to win? Trying to do what we did in the previous section, let us denote the two possible pure strategies (show one finger or two fingers) by 1 and 2. The payoff matrix for Player *I* is:

		II		
		1	2	*min*
I	1	\$1	−\$1	−\$1
	2	−\$1	\$1	−\$1
	max	\$1	\$1	

As we can see, the lower value of the game is −\$1 and it is different from the upper value of the game which is equal to \$1. There is no equilibrium point. Indeed, if Player *I* chooses the strategy 1, which is one of the two maximin strategies, then the outcome of the game is not predictable; it could be either \$1 or −\$1, depending on Player *II*'s choice. Obviously, if Player *II* knows the intention of Player *I*, then she will choose strategy 2, winning \$1. If there is no equilibrium point, each player can take advantage of any information about what the opponent is going to do. In order to prevent this to happen, Player *I* has to play randomly. As the game is symmetric, Player *II* has to do the same. In this case, due again to the symmetry of the game, the best way of acting is to create maximum uncertainty, i.e., the maximum surprise, to the opponent by playing pure strategies 1 and 2 with equal probabilities in a repeated game. The uniform distribution:

$$\begin{pmatrix} 1 & 2 \\ \frac{1}{2} & \frac{1}{2} \end{pmatrix}$$

is the optimum random strategy for both players. The expected payoff of Player *I*, when the two-finger morra game is played repeatedly, is:

$$\$1 \times \frac{1}{2} + (-\$1) \times \frac{1}{2} = \$0$$

which is said to be a *fair* game. Practically, the game is played correctly if each player tosses a fair coin and will show one finger or two fingers if the Head or the Tail comes up, respectively.

As mentioned before, not every two-person zero-sum game has an equilibrium point in terms of pure strategies, like in the first example given above but, according to the famous minimax theorem of John von Neumann, any two-person zero-sum game has an equilibrium point in terms of mixed, or random, strategies, as it happened in our second example. The concept of probability had to be invented at least for making any such game solvable.

The paradox of rational decision making.

There are two boxes B_1 and B_2. A decision maker has to choose one of the following two actions: a_1: "Open only box B_1 and keep the content of the box."; a_2: "Open both boxes and keep their contents." A hidden robot always puts \$1,000 in box B_2 whereas it sometimes puts \$100,000 in box B_1 and sometimes not. Denote by θ_1: "The robot puts money in box B_1," and by θ_2: "The robot does not put money in box B_1." In decision theory, θ_1 and θ_2 are called states of nature. The past experience has shown that with high probability (say 0.9) box B_1 contained \$100,000 if the decision maker chose action a_1, whereas the probability of finding \$100,000 in box B_1, given action a_2 of the decision maker, proved to be very small, namely 0.05. What action to take?

The following decision table summarizes the data of the problem:

	θ_1	θ_2
a_1	\$100,000	\$0
a_2	\$101,000	\$1,000

Let us put ourselves in the position of the decision maker. There are two different ways of looking at this decision problem:

(a) *The rational viewpoint:* The robot has already acted. Either it has put money in both boxes or it has put money in just box B_2. In the first case, by opening both boxes, I shall win \$101,000. In the second alternative, if I open both boxes, I shall win $1,000 anyway. On the contrary, if I open only box B_1, I shall get $100,000 in the first alternative and nothing in the second one. Even if the robot is acting against me, trying to minimize my win allowed by the rules of the game, I shall do my best acting in order to maximize this minimum. This analysis is summarized in the following table:

	θ_1	θ_2	*min*
a_1	$100,000	$0	$0
a_2	$101,000	$1,000	$1,000

Therefore, according to the *maximin* strategy, borrowed from game theory, the best action is a_2, which means to open both boxes and collect their contents.

(b) *The irrational viewpoint*: Looking into what happened in the past, I realize that the robot acted as it were aware of my own way of acting. Almost always, or more exactly in 90% of the former cases, when I chose action a_1, deciding to open only box B_1, the robot put $100,000 in box B_1 whereas almost always, or more exactly in 95% of the former cases, when I chose action a_2, the robot put no money in box B_1. Based on the past experience, I suspect that the robot has mysterious powers predicting my way of acting and penalizing me every time I greedily decide to open both boxes. Mathematically, when my action was a_1, the mean (or average) win was:

$$E[\text{ win }|a_1] = \$100,000 \times 0.9 + \$0 \times 0.1 = \$90,000$$

whereas, when my action was a_2, the mean (or average) win was:

$$E[\text{ win }|a_2] = \$101,000 \times 0.05 + \$1,000 \times 0.95 = \$6,000.$$

Therefore, taking into account what happened in the past and giving preference to the action for which the expected payoff is maximum, the best action is a_1 which means to open only box B_1.

We have just presented two criteria, both justified, but inducing different optimum ways of acting. How would you act? What makes decision making difficult is the multitude of different criteria that might be used for selecting the "best" way of acting. A decision maker cannot be simply called "rational" or "irrational" but rather "rational with respect to a specific way of selecting the best action".

Prisoner's dilemma.

A murder is committed and two suspects are arrested on the spot. Both have a gun and from each a bullet is missing, identical with the one found in the body of the victim. There are no other witnesses. The suspects are taken to the police and put in separate cells. On the next day, they are going to appear, individually, in front of the district attorney. Both know the consequences of their future way of acting. If one remains silent and the other one talks ("I am not guilty, the other one is!"), then the one who talks will become free and the other one will go to jail for twenty years. If they both talk, then they will both get five years in jail. Finally, if they both remain silent, then, due to lack of serious evidence, each will get a year in jail for not having a gun permit. The prisoners have one night to analyse the options open to them and to make up their minds about how to act, individually, in front of the district attorney on the next day, knowing the consequences of their possible actions shown in the following table:

	Prisoner B talks	Prisoner B does not talk
Prisoner A talks	$(-5, -5)$	$(0, -20)$
Prisoner A does not talk	$(-20, 0)$	$(-1, -1)$

As the data values are symmetric, it does not really matter who is labeled A and who is labeled B. The entries from the above table referring to Prisoner A and the worst output, i.e., the maximum number of years in jail imposed on Prisoner A by Prisoner B's way of acting, are:

	Prisoner B talks	Prisoner B does not talk	min
Prisoner A talks	-5	0	-5
Prisoner A does not talk	-20	-1	-20

Obviously, Prisoner A will try to minimize the number of years in jail, or to maximize the minimum shown in the last column of the tableau, by selecting the action *talk*. His rational way of reasoning would be: "The final outcome depends on the way of acting of both prisoners. Whatever Prisoner B does, I shall do better talking. For if Prisoner B remains silent and I talk, I shall get free; whereas if Prisoner B talks, then I shall do much better by talking (getting five years in jail) than by remaining silent (getting twenty years in jail)." Such a *minimax* strategy is a sound and prudent way of individually acting against a malevolent opponent: act in order to minimize the maximum loss inflicted by the opponent's action. Due to the symmetry of the problem, Prisoner B is in a similar position and, if rational, he arrives to the same best decision, namely *talk*. But then, they will go to jail for five years whereas, if they were both silent, then they would go to prison for only one year. It is paradoxical to see two brutes, unable to open their mouths, doing better than two rational beings acting according to the sound minimax strategy. Obviously, if the two prisoners could communicate and analyse the problem together, they would realize that remaining silent would be a better solution for both. But in such a case, dialog and mutual trust would play an essential part. Imagine what will happen if on the next day, facing the district attorney individually, one of the prisoners sticks to the best collective action, remaining silent, and the other one does not!

Generally, a pair of pure strategies of the two players is an equilibrium point of the game if neither player can do better by unilaterally abandoning his/her strategy. In the example just given (*talk,talk*) is an equilibrium point of the game and its consequence is $(-5, -5)$, i.e., five years in jail for each player. Such an equilibrium point is the optimum individual solution of the game. There is a better solution for both players, the optimum cooperative solution (*do not talk,do not talk*), giving the outcome $(-1, -1)$, i.e., only one year in jail for each player, but the implementation of such a cooperative solution is based on dialog and trust.

A generalization of this game is shown in the tableau given below, where NC means the noncooperative strategy, C the cooperative strategy, p is the penalty for noncooperating, r the reward for cooperating, t the temptation of double-crossing, and s the payoff of the double-crossed player:

	NC	C
NC	(p,p)	(t,s)
C	(s,t)	(r,r)

Such a game is a prisoner's dilemma type of game if $t > r > p > s$. In the previous example, we had the values: $t = 0, r = -1, p = -5, s = -20$.

The paradox of evidence.

Many of our beliefs and actions are based on evidence: either current new available evidence, or evidence from our past experience, or evidence provided by other sources. Choosing a certain doctor, taking a course from a certain professor, deciding to see a certain movie, buying a certain book, or going to a certain restaurant, we rely on our knowledge acquired from current advertisments, past experience, or other people's opinions and testimonies. A verdict is reached by a jury in a court after examining circumstantial evidence and listening to testimonies of some reliable or less reliable witnesses. The quality of evidence is not the same: Some evidence is stronger, other evidence is weaker. A body of evidence may confirm or infirm a certain hypothesis or may have no influence at all on our beliefs. A new evidence does not prove that a certain hypothesis is true but can prove its falsehood. If the body of evidence confirms an hypothesis, our belief in that hypothesis increases with a certain higher or lower degree.

Two hypotheses H_1 and H_2 are equivalent if they are simultaneously true or simultaneously false. It is generally accepted that if two hypotheses are equivalent, then any evidence that confirms one of them confirms the other one as well. Let us consider H_1: "All horses have legs," and H_2: "Everything without legs is not a horse." Obviously, H_1 and H_2 are equivalent hypotheses. If we see a horse and we notice that it has legs, then this new evidence does not prove H_1 but only confirms it. Let us assume now that the new evidence consists of a ball. It confirms H_2 but it would be absurd to say that it confirms H_1 too; it says nothing relevant about the first hypothesis. To give another example, let us take the hypothesis H_3: "All snakes live outside Newfoundland." Finding a snake outside Newfoundland should be a new evidence confirming H_3 but in fact if more and more snakes are found outside Newfoundland it will become less and less probable to believe that Newfoundland could be snakefree.

We generally give more significance to coincidence and improbability whereas the bulk of false predictions are conveniently ignored and eventually forgotten.

Surprise in conflicts.

"It is pardonable to be defeated, but never to be surprised" – is a statement attributed to Frederick the Great. History of humanity is full of examples in which countries or groups of countries have played dangerous games trying either to take advantage of the achievement of surprise or to avoid becoming the victims of surprise.

1.8. Uncertainty

Janus is a Roman god, the patron of the beginning of the day, month, and year. The first month of the year is named after him. In art, he is represented with two faces looking in opposite directions, forward and backwards. As ordinary human beings, we have only one face and, as somebody put it, we are advancing with our back to the future, facing only the past. We should like to know the future but we can only learn from the past, trying nevertheless to predict future events. Are we able to accurately forecast what is going to happen? To some extent yes, and this is why we are surviving, but possible surprises of all kinds make us live with uncertainties for ever.

We have a passion for certainty and objective. What can, however, be coonsidered to be entirely objective and certain? No observation or theory can see the whole and therefore can be absolutely right. Nature is full of uncertainties. Surprise, a key concept in the prediction paradox, is directly related to uncertainty. Roughly speaking, surprise has to do with not knowing beforehand and, consequently, it may be related to some kind of probabilistic experiment. In the finite case, a *probabilistic experiment A* is completely characterized by its possible *outcomes*, called also *elementary events*, denoted by $a_1, \ldots, a_m$, and their *probabilities* denoted by $\xi_1, \ldots, \xi_m$. Before performing the experiment, we cannot predict with certainty the future outcome. All we can say is that there is a probability ξ_i that outcome

a_i is going to occur. The elementary events are *mutually exclusive*: two outcomes of the same probabilistic experiment cannot occur simultaneously. A convenient way of representing a probabilistic experiment A is by using the array:

$$A = \begin{pmatrix} a_1 & \cdots & a_m \\ \xi_1 & \cdots & \xi_m \end{pmatrix}.$$

Thus, rolling a fair die may be viewed as a probabilistic experiment with six possible mutually exclusive outcomes, labeled 1,2,3,4,5,6 and the corresponding probability of each outcome equal to 1/6. The corresponding array would be:

$$A = \begin{pmatrix} a_1 & a_2 & a_3 & a_4 & a_5 & a_6 \\ \frac{1}{6} & \frac{1}{6} & \frac{1}{6} & \frac{1}{6} & \frac{1}{6} & \frac{1}{6} \end{pmatrix}.$$

Dealing with the possible outcomes of a probabilistic experiment, we must notice their unity in spite of their diversity. Obviously, the possible outcomes are different in some respect and this is why we may put different labels on them, but, at the same time, they have something in common, namely, they are elementary events of the same kind, being possible outcomes of the same probabilistic experiment.

Before performing a probabilistic experiment, the observer is uncertain about the forthcoming outcome, being able to predict only with probability ξ_i that the outcome will be a_i. *After* performing the probabilistic experiment, the observer is surprised by its outcome. The posterior surprise supplied by the probabilistic experiment is measured by the prior uncertainty about its forthcoming outcome. The larger the prior uncertainty the larger the posterior surprise.

Quantitatively, the amount of uncertainty contained by the experiment A is measured by the entropy:

$$H(A) = -\xi_1 \log \xi_1 - \ldots - \xi_m \log \xi_m.$$

C. Shannon (1948) introduced it by analogy with a similar formula used by L. Boltzmann in his studies of statistical mechanics. It is easy to prove that $0 \le H(A) \le \log m$. The maximum value, namely, $H(A) = \log m$, occurs if and only if all the outcomes of A are equally likely, i.e.,

$$\xi_1 = 1/m, \ldots, \xi_m = 1/m,$$

which corresponds to maximum uncertainty on the possible outcomes of the experiment A. At the same time, if the experiment A has only one possible outcome, say a_1, then:

$$\xi_1 = 1, \xi_2 = 0, \ldots, \xi_m = 0,$$

and the corresponding entropy $H(A) = 0$. Indeed, such an experiment contains no uncertainty at all, the occurrence of the outcome a_1 being certain.

> **Claude E. Shannon** (1916-2001). American engineer and mathematician. Born at Petoskey, Michigan. Graduated at the University of Michigan in 1936. Became a research assistant at the electrical engineering department of the Massachusetts Institute of Technology, in 1936, and a permanent member of the faculty there, in 1958. He got a PhD in mathematics at MIT in 1940. His master's thesis, *A symbolic analysis of relay and switching circuits*, published in 1938, applied the Boolean algebra to the study of automatic electric circuits. Creator of the mathematical information theory in 1948.

Apparently, some ancient Greek philosophers expressed the thought that natural laws prevail through an enormous number of random events. The theory of probability, however, began with the correspondence between Blaise Pascal (1623-1662) and Pierre Fermat (1601-1665) in 1654 about games of chance even if the notion of probability did not occur explicitly in their letters. In 1658, Christian Huyghens published a book entitled *De Ratiociniis in Aleae Ludo* (*Reasoning on Games of Chance*) where not probability but rather expectation, defined as the value of the hope, was taken as the basic concept. According to Huyghens, "to have ξ chances of obtaining a, and η chances of obtaining b, chances being equal, is worth $(\xi a + \eta b)/(\xi + \eta)$." The seminal book *Ars conjectandi* (*The Art of Conjecture*) by Jacob Bernoulli, appeared in 1713, eight years after the death of its author, contained the first definition of probability as "the degree of certainty, which is to the certainty as the part to the whole." The important study of Abraham de Moivre, *De Mensura Sortis, seu de Probabilitate Eventuum in Ludis a Casu Fortuito Pendentibus* (*On the Measure of Chance or on the Probability of the Results in Games of Chance*), published in 1711, and mainly the fundamental work of Pierre-Simon de Laplace (1749-1827), *Essai philosophique sur les probabilités* (*Philosophical Essay on Probabilities*) and the book *Théorie analytique de la probabilité* (*Analytical Theory of Probability*), summarized the results of the classical probability theory and gave a decisive impulse to its further development. Refining the concepts introduced by Bernoulli, Laplace gave the classical definition of probability in the following formulation: "The probability of an event is the proportion of the number of favourable cases to the number of all possible cases, where the individual cases are supposed to be equally possible."

Besides the questions of games of chance there were problems about mortality tables and insurance problems which were dealt with at the beginning of the probability theory. Using the regular weekly registering of deaths began in London in 1592, John Graunt calculated in 1662 the probability of mortality as a function of age. In Holland, Jan Hudde and Jan de Witt applied such probabilities to the calculation of life annuities.

In the nineteenth century, C.F. Gauss, P.-S. Laplace, S.-D. Poisson, P. Chebyshev, and A. Markov obtained new results in the theory of probability but the rigorous analysis of its axiomatics had to wait for the twentieth century. In 1919, Richard von Mises (1883-1953) defined probability of an event as being the limit of the relative frequences of the occurrences of this event when the number of trials increases indefinitely. If an experiment is repeated under identical conditions n times and a given event occurs m times, $(m \leq n)$, then the relative frequency of this event in the n repetitions of the experiment is m/n. The strong point in von Mises's definition consisted in relating probability, as a theoretical concept, to relative frequency, which is an experimental concept. The objectionable part of it referred to the limiting process required by his definition: an experiment cannot be repeated indefinitely and the relative frequencies do not generally form a convergent sequence of numbers; therefore we cannot talk about the limit of an infinite sequence of relative frequencies.

In 1933, A.N. Kolmogorov published the axiomatic of probability theory based on a measure-theoretical basis in a small but famous book *Grundbegriffe der Wahrscheinlichkeitsrechnung*. According to the law of large numbers, the probability of an event is viewed as being a fixed number around which the relative frequency fluctuates in an unpredictable manner, but from which it deviates at most only a little during all these random fluctuations. If we increase the number of observations, the deviation of the frequency from the probability will mostly decrease. While the probability of a random event is a definite (though sometimes not exactly known to us) number which does not depend on chance, the relative frequency of an event is an uncertain, chance-dependent number, the exact value of which cannot be determined in advance. It is true that when the experiment is repetead n times then the absolute frequency of an event is a number accessible to us by simply counting the number of occurrences of the respective event and the relative frequency is effectively obtained by dividing the absolute frequency by n. If, however, we take another series of n repetitions of the same experiment than the corresponding value of the relative frequency of the same event could be a different number. If we know the probability, then we can foresee more or less exactly the value of the relative frequency. If the probability is not known, then from the calculation of the relative frequency we can conclude the value of the probability with more or less precision. The evaluation of the relative frequency, as any measuring procedure, provides only an inexact value for the corresponding probability but the inexactness of the measurement can be decreased by increasing the number of observations. The modern textbooks on probability theory use the axioms given by Kolmogorov (1950), which rather refer to an absolute probability measure. Let us mention, however, that A. Rényi (1955) constructed a system of axioms for conditional probability, which is a relative probability measure depending on the context.

Apparently, Aristotle regarded the unexpected and the novel as the causes of surprise. The Stoics observed the connection of unexpectedness to surprise when they insisted that the wise man was astonished at nothing. The Sceptics regarded the rare, the sudden, the unusual, and the novel, as the causes of surprise. For Descartes surprise is caused by the sudden, unexpected arrival of an impression. Some psychologists emphasize that surprise occurs when the stimulus is both novel and unexpected but there are cases where surprise occurs even though the stimulus is familiar or known. It is debatable whether a stimulus causes surprise because of its novelty, or its unexpectedness, or its suddenness. Darwin believes that the unknown as well as the unexpected can cause surprise. There is both an objective and a subjective component of surprise. The occurrence of a rare or unexpected event is generally objective, whereas the significance of the surprising event is highly subjective. Surprise reflects a kind of conflict between a present, actual experience and different expectations in the mind based on past experience. Such a conflict could be minor or major, in the last case it could ask us to adapt ourselves to the new and unexpected circumstances. Surprise could be positive or negative for the subject experiencing it. It could generate emotions like fear, wonder, or curiosity. There is some indication that the surprising event may be easily remembered due to the heightened attention paid to it. Psychologists have noticed the inhibitory effect of surprise on the activity in which the individual is engaged at the moment of the experience. The intensity of emotions is increased when they are associated with surprise. The stimulus that is sudden and unexpected would produce more marked effects than one which is gradual and expected, enhancing the intensity of the emotions associated with it. Surprise tends to occur when a conscious or nonconscious expectation is frustrated. A stimulus fails to evoke surprise if it corresponds precisely to what is expected, fulfilling an expectation. Not every unexpected stimulus, however, is surprising. Surprise is caused when there is a conflict or incompatibility between the unexpected stimulus and some nonconscious expectation, knowledge, belief, or previous experience. Novelty is not an indispensable characteristic of the stimulus of surprise, although the new may arouse surprise. Surprise occurs because a present expectation is frustrated. Even known familiar outcomes may cause surprise if they occur ib hitherto uncognized relationships. Thus, newness of relationships does seem to play an important role in the causation of surprise. Therefore, surprise tends to occur when a conscious or nonconscious expectation is frustrated or when the stimulus conflicts with or is incompatible with previous knowledge, belief, or experience. Monte Carlo fallacy: the longer the run of successive reds on a fair roulette wheel the less likely it is that red will come up on the next spin.

The paradox of the heap.

If we have a heap of sand and one grain of sand is taken away, then what remains is still a heap. Therefore, all collections of grains of sand, even those consisting of one grain of sand, are heaps. The paradoxical argument here is induced by the vagueness of the word "heap". There is no precise number of grains that is enough for a collection to be a heap, and below which collections are not heaps. As we frequently use vague words – like tall, bald, or intelligent, for instance – similar paradoxes may be easily formulated. Vagueness is a main feature of our thought. It is an useful tool in coping with complexity; ignoring cumbersome details it allows us to group in the same class entities that otherwise differ in many respects. Vagueness must be distinguished from ambiguity induced either by homonyms or by lack of knowledge.

The baldness paradox.

A man is either bald or not. How many hairs is a man supposed to have in order to be classified as bald? Classic logic assumes the existence of a clean-cut answer to having or not a certain property. The answer to the last question, however, does not belong to such a category and there are many other questions like it. In order to cope with such cases, L. Zadeh (1965) has developed the so called *fuzzy logic*. Instead of saying that a man is bald or not which would be equivalent to saying that for any man the sentence "This man is bald" is either true or false, we may introduce several degrees of truth, in fact any number between 0 (false) and 1 (true). In such a case any man is bald with a certain degree $r, (0 \leq r \leq 1)$, or not bald with the degree $1 - r$. The classic logic is obtained as a particular case when the degree of truth r may take only two values 0 or 1. In a fuzzy logic the self-referential paradoxes disappear because the same sentence could be both true and false with a certain degree. Therefore, the strict alternative "true or false but not both", from the classic logic, is replaced by "true and false with certain degrees" in the fuzzy logic.

Everyday life.

In a rainy place people spend more time in pubs. Quite often, the customers discuss, debate, dispute, and contradict each other about the most unusual and extraordinary things, facts, happenings, and events. On September 12, 1954, Sir Hugh Beaver invited Norris and Ross McWhirter to see if their fact and figure agency in London could help settle arguments about records. Consequently, at the end of August 1955, the first Guinness Book of Records was published in a 198-page edition. In less than half a year it became a bestseller. Thirty years later it earned the reputation of being the top-selling copyright book in publishing history.

Many people have wondered for a very long time why is our world soccer mad? The sociologists could not come up with a unique answer to this question but, without any doubt, soccer is the most unpredictable sport, being played with the feet, much less perfect that the human hands. A soccer game contains a large amount of uncertainty about the outcome of each action and a weaker and dominated team can still win a match in the end. Who can forget the final game of the 2000's Europeean Soccer Championship when Italy lost the title to France in the last seconds of the injury time?

In a study about surprise, M. Desai (1939) made a nice association of surprise with conditions of unpreparedness. Emotions seem to commence with surprise and have the tendency to balk impulses. The thieves know this very well when they spill some ketchup on the clothes of Japanese tourists before stealing their attaché case or when they give a little push to distract the attention of the innocent pedestrian in a crowded area while taking his purse. "Since a dim, indeterminate awareness is probably the first experience in ontogeny"– Desai writes – "it is not to be wondered at that the first emotion may be of the nature of surprise. Perhaps surprise appeared very early in the evolutionary scale. At the dawn of life, in the first unicellular organism which had just emerged from lifelessness, perhaps the slightest change in the environment led to complete immobility, a relapse into lifelessness ... [as a response to] the awareness of change. Later on, perhaps, the organism learnt to recommence its activity in the face of change. Cessation of its own activity in the interests of a reorientation of the organism, followed by a readjustment, in response to a change in its environment, was perhaps one of the earliest characteristics of life. That primary form of response probably persists all through the evolutionary scale." The unexpectedness of life itself continuously generates all sorts of positive and negative surprises.

As mentioned before, according to Frederick the Great, it is pardonable to be defeated but never to be surprised. Human relations have been strongly influenced by the logic of surprise: exploit the advantages generated by the achievement of surprise and avert becoming the victim of surprise. Often the aggressor uses deception to surprise the victim and the surprise is attempted during a period in which the two parties are on friendly terms with one another. A couple of years ago, after a long transatlantic flight from Toronto, I arrived in Amsterdam, together with my wife, very early in the morning. Taking the train from Schiphol airport to the Central Station, I was supposed to spend some days in Amsterdam whereas my wife wanted to continue the trip by train to Erlangen in Germany. Taking a short break we stopped for a couple of minutes. At that early time of the morning not many people were around. Suddenly, a teenager, looking a little confused, approached me asking for some directions about how to get to Utrecht. I put my luggage in front of my wife and willing to help the young fellow I moved some steps closer to a train timetable. My wife also moved closer to assist herself. All explanations lasted a minute or two. I could swear that nobody was around but we three. The teenager disappeared hurrying to catch the first train to Utrecht but so did the most important piece of my luggage, a bag with everything important in it, substituted by an accomplice, perhaps hidden behind a thick column. It was like magic. Instantly, I could not realize what really had happened and I wasted precious moments being stunned and unable to do anything. At the police station, a very polite officer could only wish us to have a nice day and a pleasant stay in Amsterdam. Of course, the mistake was mine. After so many good things learnt about how honest people in Europe are, I was not prepared to avert the unpleasant surprise.

The decision to resort to surprise is not always a rational one and sometimes it underrates the risks involved. If winning a war is the main objective, then Japan's decision to surprisingly attack Pearl Harbor in 1941 lacked rationality since the United States was stronger. If, as a matter of pride, winning a certain battle is valued more highly than winning the war, then Japan's decision can no longer be perceived as irrational. Whatever the real reasons were, on December 7, 1941, after months of arduous preparation for war, Japan launched a surprise attack on Pearl Harbor. The awful and tragic terrorist attack in New York and Washington on September 11, 2001, was a big negative surprise, but an even bigger surprise was the fact that the terrorists used hijacked American planes, full of fuel, as their deadly weapons.

There are many different definitions of what a chemical, cybernetic, biological, social, or political system really means. In spite of many definitions given by different specialists from different fields, there is a common acceptance that a system of any nature represents something more than the simple juxtaposition of its parts. The connection between systems and subsystems, at different levels of reality, is universal, one of the most important charactersitics of our world. There is no exageration in saying that it is impossible to lift a finger without troubling a star. This "something more" mainly consists of the inner interdependence among the components of the system and the external interdependence with other systems. Very often, a system may be observed, examined, and explained only by isolating it, ignoring its ties with other systems. There are no independent systems but, in our struggle of coping with complexity, we prefer to cut down, neglect, or ignore weak internal or external interdependences. The human body is an example of complex system which cannot function without the interdependence between its components. It is remarkable that the informational entropy, introduced as a measure of the amount of uncertainty contained by a probabilistic experiment, has been subsequently used for measuring the amount of interdependence among the subsystems of a system and the total connection between the components of a system. Thus, the difference between the sum of the entropies of the subsystems making up a system and the entropy of the system itself is an excellent measure of the amount of interdependence between these subsystems. Also, the difference between the sum of the entropies of all the components of a system and the entropy of the system itself measures the total amount of internal connection of the respective system.

1.9. Conclusion

Originally, logic dealt with the external forms of reasoning, often empty of content. The logic was formal and emphasized what is stable in our way of manipulating two-valued (i.e., true-false) sentences by using elementary operations, like disjunction (or), conjunction (and), and negation (non), and fundamental properties involving these operations, like the principle of contradiction, the principle of excluded third, and the rules of correct negation known as De Morgan's rules. The old classic logic has been changed and generalized by all sorts of nonstandard logics including dialectic logic, modal logics, multi-valued logics, and, more recently, the fuzzy logic. A new logic has been and is being built with the ambitious aim of mathematically catching the internal subtleties of human reasoning, with its many nuances. Logic is asked to explain change, transformation, development, flexibility, uncertainty, surprise, novelty, and general connection. Only such a logic could be implemented in order to build future generations of computers which would be not only faster than the human beings, as they already are, but at least as intelligent as the brightest human beings.

The present monograph does not contain such an absolute and ultimate Logic, with capital letter L. Its aim is only to formalize a relative variant of logic which is probabilistic, temporal, and essentially depends on circumstances or/and a specific context. It has been repeatedly said that the old principle of excluded third, known as *tertium non datur*, stating that a sentence is either true or false, but not both, has been infirmed by the apparition of the multi-valued logics where sentences are allowed more than two logical values. It has also been said that in its ontological form, i.e., "Everything is or is not," the principle of excluded third, which is equivalent with the ontological form of the principle of contradiction, i.e., "There exists nothing that is and is not," must be valid in any kind of logic. This is indeed so if time is frozen and circumstances are kept strictly unchanged or are simply ignored. In the relative logic, however, something can exist now and cannot exist later, or can exist or not even now, at the same time, if the context and circumstances are different. And such things could occur with certainty or with some probabilities depending on time and circumstances or context. The classic logic is rigid and categoric: the sentences p and non-p cannot be both true. The relative logic is flexible. Instead of p and non-p both true, p and non-p may be both possible, or they may be both true at different instants of time or under different circumstances.

Context is everything. Many paradoxes from classic logic simply disappear in the relative logic. Going back to the barber paradox, we remember that the regiment's barber shaves all and only those who do not shave themselves. The dilemma is about what does happen with the barber himself? The barber shaves himself if and only if he does not. There is no such paradox if we define the regiment's barber to be the member of the regiment who shaves all and only the members of the regiment who do not shave themselves before 10 a.m. There is now no ambiguity about what the barber should do with himself; he simply has to postpone his shaving until sometime after 10 a.m. on the same day.

The jokes are generally based on the possibility of looking at things and facts from different angles. When the innocent question: "Are dead people buried with a priest in your village?" is answered: "No, in our village the priest remains outside!", the comic simply comes from the fact that the question may be looked upon from two different contexts and the answer chooses the least expected one. The language is full of ambiguity when the context is not carefully defined or explained. In a mystery novel the set of possible suspects is changing according to the different circumstances and relevant or irrelevant evidence gathered or revealed to us by the author during the action. In the end, only Hercule Poirot succeeds in revealing to us the real factual context which unambiguously explains who the culprit is.

The classic logic has completely ignored uncertainty in its formalism. In fact, applying correctly its rules was seen as a method of eliminating uncertainty and ambiguity from our way of thinking and behaving. However, uncertainty is a fact of life and we are facing it all the time in analysing different possibilities and alternatives. Even in an ideal world, governed by strictly deterministic laws, as anything is connected to everything, we cannot grasp the entire complexity and, inevitably, we ignore some conditions, some factors, or/and some causes. Uncertainty is generated in such a case by not knowing everything which would make the events fully predictable. In a real world, uncertainty is an objective part of it and the random fluctuations do occur, whether we like them or not. In any realistic prediction or forecasting model, we want to find a trend and, maybe, some kind of periodicity, but the random fluctuations are inevitable and must be included into such a model.

The minimax theorem tells us that in a two-person zero-sum game, which is a model which describes the competition between two rational individuals with opposite interests, where the gain of one player is equal to the loss of the opponent, the optimum solution means, more often than not, to create uncertainty by using a mixed strategy, which is a random way of alternating the possible pure strategies. But this is only a very simple model because we are not always fully rational and have to make decisions in a world endowed neither with perfect knowledge nor perfect competition. We also behave in groups, with common interests, and there are better cooperative solutions than the individual ones. Only a relative logic, which is a probabilistic, temporal, and context-dependent logic, could cope with complexity.

In the middle of Malign Lake, in the Canadian Rocky Mountains, there is a small and mysterious island, called the Spirit Island, where the legend says that the chiefs of different Indian tribes used to meet and discuss before taking important decisions. These chiefs were always old, their judgements being based on a way of thinking heavily relying on a long life experience, when time and different circumstances helped them to become wiser. The natural history, the history of humanity, the passage of time, and the continuous modification of the circumstances can only teach us that the problem with truth is its many varieties.

Chapter 2. TWO-VALUED RELATIVE LOGIC

2.1. Definitions and notations

The pair of ordered components a and b: (a, b);

The vector of ordered components $a_1, \ldots, a_n$: $(a_1, \ldots, a_n)$;

The set consisting of the elements $a, b, c, \ldots$: $\{a, b, c, \ldots\}$;

The set of real numbers from the unit interval: $[0, 1]$;

The set of binary $0 - 1$ vectors with n components: $\{0, 1\}^n$;

The set of real numbers: $\mathbf{R}$;

Truth values: $k, k_1, k_2, \ldots$ from $\{0, 1\}$, where $0 = $ *false* and $1 = $ *true*;

Sometimes the logical values "true" and "false" are denoted also by T and F, respectively;

The element a belongs to the set A: $a \in A$;

The element b does not belong to the set A: $b \notin A$;

The complementary set: $\bar{A} = \{x; x \notin A\}$;

The union of two sets: $A \cup B = \{x; x \in A \text{ or } x \in B\}$;

The intersection of two sets: $A \cap B = \{x; x \in A \text{ and } x \in B\}$;

The difference of two sets: $A - B = \{x; x \in A \text{ and } x \notin B\}$;

The inclusion of a set A in the set B: $A \subseteq B$, which means that every element of A belongs also to B;

The strict inclusion of a set A in the set B: $A \subset B$, which means that every element of A belongs to B but there is at least one elemnt from B which does not belong to A;

The equality of two sets: $A = B$, which means that A and B contain the same elements;

The product set of the sets A and B: $A \times B = \{(a, b); a \in A, b \in B\}$;

The class of all subsets of a set X: $\mathcal{P}(X) = \{A; A \subseteq X\}$;

The conjunction "and" is represented by a dot: $\cdot$;

The disjunction "or" is represented by $\vee$;

The implication: $\Rightarrow$;

The equivalence: $\Leftrightarrow$;

The negation: $\sim$;

Atomic (simple) sentences: $p, p_1, p_2, \ldots$;

Molecular (complex) sentences: $P, P_1, P_2, \ldots$;

Circumstances: $Q, Q_1, Q_2, \ldots$;
Time instants: $t, t_1, t_2, \ldots$;

2.2. Classic logic

Aristotle's traditional logic was a logic of concepts and of relationships between concepts. The Stoics, especially Chrysippos, considered sentence as fundamental concept of classic logic, starting from the hypothetical syllogism:

Modus ponens	*Modus tollens*
If p_1 is, p_2 is.	If p_1 is, p_2 is.
But p_1 is.	But p_2 is not.
Then p_2 is.	Then p_1 is not.

where p_1 and p_2 are sentences. Whitehead and Russell (1925) used the same idea in rigorously elaborating the calculus of classic logic.

Sentences are classified in atomic or elementary sentences, like "Snow is white," and molecular sentences consisting of elementary sentences combined together by using connectors like "and", "or", "if ... then". A sentence in classic logic may be "true", denoted by 1 or T, or "false", denoted by 0 or F. The operations on sentences and the common symbols currently used in classic logic are:

Negation: $\sim p$,   "non-p" or "not p";
Disjunction: $p_1 \vee p_2$,   "p_1 or p_2";
Definition: $=$,   (to define a molecular sentence);
Conjunction: $p_1 \cdot p_2$,   "p_1 and p_2";
Implication: $p_1 \supset p_2$,   "p_1 implies p_2";
Equivalence: $p_1 \equiv p_2$,   "p_1 is equivalent to p_2" (p_1 and p_2 have the same truth value; they are simultaneously true or false);
Incompatibility: $p_1 \div p_2$,   "p_1 is incompatible with p_2" (p_1 and p_2 cannot be simultaneously true, therefore at least one of them is false);
Parentheses: $(\ldots)$,   (to show the order of operations);
Assertion symbol: $\top p$,   "p is true, or p is affirmed"; $\top(p_1 \supset p_2)$,   "it is true that p_1 implies p_2" (here the implication is affirmed whereas sentences p_1 and p_2 are only considered; we affirm nothing about them);

Inference:

Modus ponens	Modus tollens
$\top(p_1 \supset p_2)$	$\top(p_1 \supset p_2)$
$\top p_1$	$\sim \top p_2$
$\top p_2$	$\sim \top p_1$

Truth function: $f(p_1, p_2)$, (a function depending on the truth values of the argument sentences p_1, p_2; its possible values are also the logical values "true" or "false"). The implication, disjunction, conjunction, equivalence, incompatibility, are simple examples of truth functions. The truth functions were systematically studied by Wittgenstein (1933). The following tableau gives the logical values of the elementary truth functions:

p_1	p_2	$p_1 \cdot p_2$	$p_1 \vee p_2$	$p_1 \supset p_2$	$p_1 \equiv p_2$	$p_1 \div p_2$
1	1	1	1	1	1	0
0	1	0	1	1	0	1
1	0	0	1	0	0	1
0	0	0	0	1	1	1

Not all truth functions are independent. Thus, we can easily prove, by checking the logical values of the corresponding truth functions with respect to all possible logical values of their arguments, the following equalities between sentences:

$$p_1 \cdot p_2 = \, \sim (\sim p_1 \vee \sim p_2);$$

$$p_1 \supset p_2 = \, \sim p_1 \vee p_2;$$

$$\sim p = p \div p;$$

$$p_1 \vee p_2 = (p_1 \div p_1) \div (p_2 \div p_2);$$

$$p_1 \cdot p_2 = (p_1 \div p_2) \div (p_1 \div p_2);$$

$$p_1 \supset p_2 = p_1 \div (p_2 \div p_2).$$

Georg Cantor's set theory, on which modern mathematics is based (Bourbaki (1968, 1999), Bell (1965, pp.555-579)), is similar to the classic logic of concepts. If A, B are arbitrary sets of a universe X, then the complementary set:

$$\bar{A} = \{x; x \in X, x \notin A\},$$

the union:
$$A \cup B = \{x \in A \text{ or } x \in B\},$$

the intersection:
$$A \cap B = \{x \in A \text{ and } x \in B\},$$

and the inclusion:

$$A \subset B, \text{ meaning that if } x \in A \text{ then } x \in B,$$

correspond to the classic logical operators: "not" $\sim$, "or" $\vee$, "and" $\cdot$, and "implies" $\supset$.

2.3. Two-valued relative logic

Throughout this chapter we assume that each sentence can be either true (denoted by 1 or, sometimes, T) or false (denoted by 0 or, sometimes, F) but these logical (or truth) values are taken with some probabilities. The two-valued logic is binary, temporal, probabilistic, and context dependent.

The probability distribution (in the objective case) or credibility distribution (in the subjective case) at time t that the atomic sentence p_1, under circumstances Q_1, has the truth value k_1, and ... the atomic sentence p_n, under circumstances Q_n, has the truth value k_n is denoted by:

$$\varphi_t\{p_1/Q_1, \ldots, p_n/Q_n\}[k_1, \ldots, k_n] \geq 0,$$

and satisfies the equality:

$$\sum_{\{(k_1,\ldots,k_n); k_i \in \{0,1\}, (i=1,\ldots,n)\}} \varphi_t\{p_1/Q_1, \ldots, p_n/Q_n\}[k_1, \ldots, k_n] = 1.$$

Often, φ_t is called the logic probability (credibility) at time t. It is a joint probability distribution at time t:

$$\varphi_t : \{0,1\}^n \longrightarrow [0,1].$$

Even if the values taken on by the logic probability distribution φ_t are any real numbers from the unit interval $[0,1]$, we use to say that the truth (logical) value k of sentence P, under circumstances Q at time t, is "certain", "highly probable", "lowly probable", or "impossible", if:

$$\varphi_t\{P/Q\}[k] = 1, \quad 0.5 \leq \varphi_t\{P/Q\}[k] < 1,$$

$$0 < \varphi_t\{P/Q\}[k] < 0.5, \quad \varphi_t\{P/Q\}[k] = 0,$$

respectively. We write:

$$\varphi\{P/Q\}, \qquad \varphi_t\{P\}, \qquad \varphi\{P\},$$

if time t, or circumstances Q, or both time t and circumstances Q, are irrelevant or obviously understood, respectively.

2.4. Operations

There are three main logical operations: the negation, the disjunction, or logical sum, and the conjunction, or logical product.

Negation $\sim$ (*non*).

$$\varphi_t\{\sim p/Q\}[k] = \varphi_t\{(\sim p)/Q\}[k] = \varphi_t\{p/Q\}[\bar{k}],$$

where $\bar{k} = 0$, if $k = 1$, and $\bar{k} = 1$, if $k = 0$.

Disjunction (logical sum) $\vee$ (*or*).

$$\varphi_t\{p_1/Q_1 \vee p_2/Q_2\}[1] = \sum_{(k_1,k_2)\neq(0,0)} \varphi_t\{p_1/Q_1, p_2/Q_2\}[k_1, k_2],$$

$$\varphi_t\{p_1/Q_1 \vee p_2/Q_2\}[0] = \varphi_t\{p_1/Q_1, p_2/Q_2\}[0, 0],$$

referring to the values of p_1/Q_1, or p_2/Q_2, or both.

Conjunction (logical product) $\cdot$ (*and*).

$$\varphi_t\{p_1/Q_1 \cdot p_2/Q_2\}[1] = \varphi_t\{p_1/Q_1, p_2/Q_2\}[1, 1],$$

$$\varphi_t\{p_1/Q_1 \cdot p_2/Q_2\}[0] = \sum_{(k_1,k_2)\neq(1,1)} \varphi_t\{p_1/Q_1, p_2/Q_2\}[k_1, k_2].$$

2.5. Relations

Logical implication $\Rightarrow$ (logically implies).

$$\varphi_t\{p_1/Q_1 \Rightarrow p_2/Q_2\}[1] =$$

$$= \sum_{(k_1,k_2)\neq(1,0)} \varphi_t\{p_1/Q_1, p_2/Q_2\}[k_1, k_2],$$

$$\varphi_t\{p_1/Q_1 \Rightarrow p_2/Q_2\}[0] = \varphi_t\{p_1/Q_1, p_2/Q_2\}[1, 0].$$

Equivalence $\equiv$ (equivalent to).

$$\varphi_t\{p_1/Q_1 \equiv p_2/Q_2\}[1] =$$

$$= \varphi_t\{p_1/Q_1, p_2/Q_2\}[1, 1] + \varphi_t\{p_1/Q_1, p_2/Q_2\}[0, 0],$$

$$\varphi_t\{p_1/Q_1 \equiv p_2/Q_2\}[0] =$$

$$= \varphi_t\{p_1/Q_1, p_2/Q_2\}[1, 0] + \varphi_t\{p_1/Q_1, p_2/Q_2\}[0, 1],$$

Absorption.

For any sentence p, circumstances Q, and time t, we have the contraction property:

$$\varphi_t\{p/Q, p/Q\}[k_1, k_2] = \begin{cases} \varphi_t\{p/Q\}[k], & \text{if } k_1 = k_2 = k, \\ \\ 0, & \text{if } k_1 \neq k_2, \end{cases}$$

showing that p/Q can take only one logical value at a given time instant t.

Logical equality (at time t).

The sentences p_1, under circumstances Q_1, and p_2, under circumstances Q_2, are logically equal, denoted by:

$$\varphi_t\{p_1/Q_1\} = \varphi_t\{p_2/Q_2\},$$

if and only if:

$$\varphi_t\{p_1/Q_1\}[k] = \varphi_t\{p_2/Q_2\}[k]$$

for all truth values k.

Independence (at time t).

The sentences p_1, under circumstances Q_1, and p_2, under circumstances Q_2, are independent at time t if:

$$\varphi_t\{p_1/Q_1, p_2/Q_2\}[k, \ell] =$$
$$= \varphi_t\{p_1/Q_1\}[k]\, \varphi_t\{p_2/Q_2\}[\ell],$$

for all truth values k, ℓ.

Total incompatibility (at time t).

The sentences p_1, under circumstances Q_1, and p_2, under circumstances Q_2, are totally incompatible at time t if:

$$\varphi_t\{p_1/Q_1, p_2/Q_2\}[k, k] = 0,$$

for all truth values k.

Total compatibility (at time t).

The sentences p_1, under circumstances Q_1, and p_2, under circumstances Q_2, are totally compatible at time t if:

$$\varphi_t\{p_1/Q_1, p_2/Q_2\}[k, \ell] - 0,$$

for all truth values k and ℓ such that $k \neq \ell$.

Remark: Two sentences, p_1, under circumstances Q_1, and p_2, under circumstances Q_2, totally compatible at time t, are equivalent with certainty at time t. Indeed,

$$\varphi_t\{p_1/Q_1 \equiv p_2/Q_2\}[1] =$$
$$\varphi_t\{p_1/Q_1, p_2/Q_2\}[1, 1] + \varphi_t\{p_1/Q_1, p_2/Q_2\}[0, 0] = 1$$

Commutativity.

$$\varphi_t\{p_1/Q_1, p_2/Q_2\}[k, \ell] = \varphi_t\{p_2/Q_2, p_1/Q_1\}[\ell, k],$$

for all truth values k and ℓ.

Marginality.

$$\varphi_t\{p_1/Q_1\}[k] = \sum_\ell \varphi_t\{p_1/Q_1, p_2/Q_2\}[k, \ell],$$

for all truth values k.

Probabilistic implication.

The conditional logic probability (credibility) distribution at time t of the truth (or logical) values of sentence p_2, under circumstances Q_2, given the truth (logical) value k of sentence p_1, under circumstances Q_1, is defined by:

$$\varphi_t\{p_2/Q_2 \mid p_1/Q_1\}[\ell \mid k] = \frac{\varphi_t\{p_1/Q_1, p_2/Q_2\}[k, \ell]}{\varphi_t\{p_1/Q_1\}[k]},$$

for all possible truth (logical) values k of sentence p_1, assuming that the denominator is different from zero. For each truth (logical) value k of sentence p_1, under circumstances or context Q_1,

$$\varphi_t\{p_2/Q_2 \mid p_1/Q_1\}[\ell \mid k]$$

is a probability (credibility) distribution on the truth (logic) values ℓ of sentence p_2, under circumstances Q_2. It characterizes the *probabilistic implication*:

$$p_2/Q_2 \mid p_1/Q_1$$

read as "sentence p_2 under circumstances (context) Q_2, given (if, conditioned by) sentence p_1 under circumstances (context) Q_1".

2.6. Inference

Substitution.

Let p_1 and p_2 be two atomic (simple) sentences under circumstances (contexts) Q_1 and Q_2, respectively, and let P be a molecular (complex) sentence involving the sentence p_1 under circumstances Q_1 at time t.

If:

$$\varphi_t\{P(p_1/Q_1)\}[1] = \alpha,$$

and:

$$\varphi\{p_1/Q_1, p_2/Q_2\}[k, \ell] = 0,$$

for all truth values k and ℓ such that $k \neq \ell$, and:

$$\varphi_t\{p_1/Q_1\}[1] = \varphi_t\{p_2/Q_2\}[1],$$

and p_1/Q_1 and p_2/Q_2 are independent of the other atomic sentences involved in P, then:

$$\varphi_t\{P(p_2/Q_2)\}[1] = \alpha.$$

According to this rule, at any time t, we can replace p_1/Q_1 by a totally compatible and logically equal p_2/Q_2 without affecting the probability of the true value of the complex sentence P which involved p_1/Q_1, provided that both p_1/Q_1 and p_2/Q_2 are independent of all the other sentences involved in the complex sentence P.

Modus ponens.

(a) Using the logical implication:
If:

$$\varphi_t\{p_1/Q_1\}[1] = \alpha,$$

and:

$$\varphi_t\{p_1/Q_1 \Rightarrow p_2/Q_2\}[1] = \beta,$$

where:

$$0 \leq \alpha \leq 1, \quad 0 \leq \beta \leq 1, \quad \alpha + \beta \geq 1,$$

then:

$$\varphi_t\{p_1/Q_1, p_2/Q_2\}[1, 1] = \alpha + \beta - 1.$$

Justification: Indeed,

$$\beta = \varphi_t\{p_1/Q_1 \Rightarrow p_2/Q_2\}[1] =$$

$$= \sum_{(k_1, k_2) \neq (1,0)} \varphi_t\{p_1/Q_1, p_2/Q_2\}[k_1, k_2],$$

whereas:

$$\alpha = \varphi_t\{p_1/Q_1\}[1] = \sum_{k_2} \varphi_t\{p_1/Q_1, p_2/Q_2\}[1, k_2].$$

Summing up these two equalities and taking into account that:

$$\sum_{k_1,k_2} \varphi_t\{p_1/Q_1, p_2/Q_2\}[k_1, k_2] = 1,$$

we obtain:

$$\alpha + \beta = \varphi_t\{p_1/Q_1, p_2/Q_2\}[1, 1] + 1. \qquad \square$$

Remark: In particular, if $\alpha = 1$ and $\beta = 1$, then we obtain:

$$\varphi_t\{p_1/Q_1, p_2/Q_2\}[1, 1] = 1,$$

which shows that if p_1/Q_1 implies p_2/Q_2, with certainty, and p_1/Q_1 is true with certainty, then p_2/Q_2 is true with certainty.

(b) Using the probabilistic implication:
If:

$$\varphi_t\{p_1/Q_1\}[1] = \alpha$$

and:

$$\varphi_t\{p_2/Q_2 \mid p_1/Q_1\}[1 \mid 1] = \beta,$$

$$\varphi_t\{p_2/Q_2 \mid p_1/Q_1\}[1 \mid 0] = \gamma$$

are given, such that:

$$0 \leq \alpha \leq 1, \quad 0 \leq \beta \leq 1, \quad 0 \leq \gamma \leq 1,$$

$$\alpha\beta + \gamma \leq 1 + \alpha\gamma,$$

then:

$$\varphi_t\{p_2/Q_2\}[1] = \alpha\beta + (1 - \alpha)\gamma.$$

Justification: According to this rule, the degrees of certitude that the sentence p_1/Q_1 and probabilistic implication $p_2/Q_2 \mid p_1/Q_1$ are true at time

t completely determine the degree of certitude that the sentence p_2/Q_2 is true at time t, by using the equality:

$$\varphi_t\{p_2/Q_2\}[1] = \varphi_t\{p_2/Q_2 \mid p_1/Q_1\}[1 \mid 1]\,\varphi_t\{p_1/Q_1\}[1]+$$

$$+\varphi_t\{p_2/Q_2 \mid p_1/Q_1\}[1 \mid 0](1 - \varphi_t\{p_1/Q_1\}[1]). \qquad \square$$

Remark: In the particular case when the time and circumstances are ignored and $\alpha = 1$ (i.e., the sentence p_1 is true with certainty) and $\beta = 1$ (i.e., the sentence p_1 implies with certainty p_2), in which case $\gamma = 0$, then the sentences p_1 and p_2 are both true with certainty.

Modus tollens.

(a) Using the logical implication:
If:

$$\varphi_t\{p_2/Q_2\}[0] = \alpha,$$

and:

$$\varphi_t\{p_1/Q_1 \Rightarrow p_2/Q_2\}[1] = \beta,$$

where:

$$0 \leq \alpha \leq 1, \quad 0 \leq \beta \leq 1, \quad \alpha + \beta \geq 1,$$

then:

$$\varphi_t\{p_1/Q_1, p_2/Q_2\}[0, 0] = \alpha + \beta - 1.$$

Justification: Indeed,

$$\beta = \varphi_t\{p_1/Q_1 \Rightarrow p_2/Q_2\}[1] =$$

$$= \sum_{(k_1,k_2)\neq(1,0)} \varphi_t\{p_1/Q_1, p_2/Q_2\}[k_1, k_2],$$

whereas:

$$\alpha = \varphi_t\{p_2/Q_2\}[0] = \sum_{k_1} \varphi_t\{p_1/Q_1, p_2/Q_2\}[k_1, 0].$$

Summing up these two equalities and taking into account that:

$$\sum_{k_1,k_2} \varphi_t\{p_1/Q_1, p_2/Q_2\}[k_1, k_2] = 1,$$

we obtain:

$$\alpha + \beta = \varphi_t\{p_1/Q_1, p_2/Q_2\}[0,0] + 1. \qquad \square$$

Remark: In particular, if $\alpha = 1$ and $\beta = 1$, then we obtain:

$$\varphi_t\{p_1/Q_1, p_2/Q_2\}[0,0] = 1,$$

which shows that if p_1/Q_1 implies p_2/Q_2, with certainty, and p_2/Q_2 is false with certainty, then p_1/Q_1 is false with certainty.

(b) Using the probabilistic implication:
If:

$$\varphi_t\{p_2/Q_2\}[0] = \alpha$$

and:

$$\varphi_t\{p_2/Q_2 \mid p_1/Q_1\}[0 \mid 1] = 1 - \beta,$$

$$\varphi_t\{p_2/Q_2 \mid p_1/Q_1\}[0 \mid 0] = \gamma$$

are given, such that either:

$$\beta + \gamma < 1 \quad \text{and} \quad \alpha \leq \gamma,$$

or:

$$\beta + \gamma > 1 \quad \text{and} \quad \alpha \geq \gamma,$$

then:

$$\varphi_t\{p_1/Q_1\}[0] = \frac{1 - (\alpha + \beta)}{1 - (\beta + \gamma)}.$$

Justification: According to this rule, the degrees of certitude that the sentence p_2/Q_2 is false and probabilistic implication $p_2/Q_2 \mid p_1/Q_1$ is true, at time t, completely determine the degree of certitude of the falsity of the sentence p_2/Q_2 at time t, by using the equality:

$$\varphi_t\{p_2/Q_2\}[0] = \varphi_t\{p_2/Q_2 \mid p_1/Q_1\}[0 \mid 1](1 - \varphi_t\{p_1/Q_1\}[0]) +$$

$$+ \varphi_t\{p_2/Q_2 \mid p_1/Q_1\}[0 \mid 0] \, \varphi_t\{p_1/Q_1\}[0]. \qquad \square$$

Again, in the particular case when the time and circumstances are ignored and $\alpha = 1$ (i.e., the sentence p_2 is false with certainty) and $\beta = 1, \gamma = 1$ (i.e., the sentence p_1 implies with certainty p_2), then the sentence p_1 is also false with certainty.

Remarks: 1) We have:

$$\varphi_t\{p/Q, p/Q\}[1,0] = 0,$$

but we can have:

$$\varphi_t\{p/Q_1, p/Q_2\}[1,0] > 0.$$

2) In relative logic, it is possible to have:

$$\textit{Thesis:} \qquad \varphi_{t_1}\{p/Q_1\}[1] > 0;$$

$$\textit{Antithesis:} \qquad \varphi_{t_2}\{\sim p/Q_2\}[1] > 0;$$

$$\textit{Synthesis:} \qquad \varphi_{t_3}\{p/Q_3, \sim p/Q_4\}[1,1] > 0,$$

for $t_1 \leq t_2 \leq t_3$.

2.7. Basic properties

Proposition 2.1 (*The law of double negation*):

$$\varphi_t\{p/Q\} = \varphi_t\{\sim (\sim p)/Q\}.$$

Proof: For every truth value k we have:

$$\varphi_t\{\sim (\sim p)/Q\}[k] = \varphi_t\{\sim p/Q\}[\bar{k}] =$$

$$= \varphi_t\{p/Q\}[\bar{\bar{k}}] = \varphi_t\{p/Q\}[k]. \quad \square$$

Proposition 2.2: *We have:*

$$\varphi_t\{p_1/Q_1 \Rightarrow p_2/Q_2\} = \varphi_t\{\sim p_1/Q_1 \vee p_2/Q_2\}.$$

Proof: For the truth value $k = 1$, we get:

$$\varphi_t\{\sim p_1/Q_1 \vee p_2/Q_2\}[1] =$$

$$= \sum_{(k_1,k_2) \neq (0,0)} \varphi_t\{\sim p_1/Q_1, p_2/Q_2\}[k_1, k_2] =$$

$$= \sum_{(k_1,k_2)\neq(0,0)} \varphi_t\{p_1/Q_1, p_2/Q_2\}[\bar{k}_1, k_2] =$$

$$= \sum_{(k_1,k_2)\neq(1,0)} \varphi_t\{p_1/Q_1, p_2/Q_2\}[k_1, k_2] =$$

$$= \varphi_t\{p_1/Q_1 \Rightarrow p_2/Q_2\}[1],$$

and for the truth value $k = 0$, we have as well:

$$\varphi_t\{\sim p_1/Q_1 \vee p_2/Q_2\}[0] = \varphi_t\{\sim p_1/Q_1, p_2/Q_2\}[0,0] =$$

$$= \varphi_t\{p_1/Q_1, p_2/Q_2\}[1,0] = \varphi_t\{p_1/Q_1 \Rightarrow p_2/Q_2\}[0]. \quad \square$$

Proposition 2.3 (*The primal generalized De Morgan law*):

$$\varphi_t\{\sim (p_1/Q_1 \vee p_2/Q_2)\} = \varphi_t\{\sim p_1/Q_1 \cdot \sim p_2/Q_2\}.$$

Proof: For the truth value $k = 1$, we have:

$$\varphi_t\{\sim (p_1/Q_1 \vee p_2/Q_2)\}[1] = \varphi_t\{p_1/Q_1 \vee p_2/Q_2\}[0] =$$

$$= \varphi_t\{p_1/Q_1, p_2/Q_2\}[0,0] = \varphi_t\{\sim p_1/Q_1, \sim p_2/Q_2\}[1,1] =$$

$$= \varphi_t\{\sim p_1/Q_1 \cdot \sim p_2/Q_2\}[1],$$

and, for the truth value $k = 0$, we have:

$$\varphi_t\{\sim (p_1/Q_1 \vee p_2/Q_2)\}[0] = \varphi_t\{p_1/Q_1 \vee p_2/Q_2\}[1] =$$

$$= \sum_{(k_1,k_2)\neq(0,0)} \varphi_t\{p_1/Q_1, p_2/Q_2\}[k_1, k_2] =$$

$$= \sum_{(k_1,k_2)\neq(0,0)} \varphi_t\{\sim p_1/Q_1, \sim p_2/Q_2\}[\bar{k}_1, \bar{k}_2] =$$

$$= \sum_{(k_1,k_2)\neq(1,1)} \varphi_t\{\sim p_1/Q_1, \sim p_2/Q_2\}[k_1, k_2] =$$

$$= \varphi_t\{\sim p_1/Q_1 \cdot \sim p_2/Q_2\}[0]. \quad \square$$

Proposition 2.4 (*The dual generalized De Morgan law*):

$$\varphi_t\{\sim (p_1/Q_1 \cdot p_2/Q_2)\} = \varphi_t\{\sim p_1/Q_1 \vee \sim p_2/Q_2\}.$$

Proof: For the truth value $k = 1$, we have:

$$\varphi_t\{\sim (p_1/Q_1 \cdot p_2/Q_2)\}[1] = \varphi_t\{p_1/Q_1 \cdot p_2/Q_2\}[0] =$$

$$= \sum_{(k_1,k_2)\neq(1,1)} \varphi_t\{p_1/Q_1, p_2/Q_2\}[k_1, k_2] =$$

$$= \sum_{(k_1,k_2)\neq(1,1)} \varphi_t\{\sim p_1/Q_1, \sim p_2/Q_2\}[\bar{k}_1, \bar{k}_2] =$$

$$= \sum_{(k_1,k_2)\neq(0,0)} \varphi_t\{\sim p_1/Q_1, \sim p_2/Q_2\}[k_1, k_2] =$$

$$= \varphi_t\{\sim p_1/Q_1 \vee \sim p_2/Q_2\}[1],$$

and, for the truth value $k = 0$, we have:

$$\varphi_t\{\sim (p_1/Q_1 \cdot p_2/Q_2)\}[0] = \varphi_t\{p_1/Q_1 \cdot p_2/Q_2\}[1] =$$

$$= \varphi_t\{p_1/Q_1, p_2/Q_2\}[1, 1] = \varphi_t\{\sim p_1/Q_1, \sim p_2/Q_2\}[0, 0] =$$

$$= \varphi_t\{\sim p_1/Q_1 \vee \sim p_2/Q_2\}[0]. \quad \square$$

Proposition 2.5 (*The principle of excluded third – tertium non datur*):

$$\varphi_t\{p/Q \vee \sim p/Q\}[1] = 1.$$

Proof: The above equality holds because, for the truth value $k = 0$, we have:

$$\varphi_t\{p/Q \vee \sim p/Q\}[0] = \varphi_t\{p/Q, \sim p/Q\}[0, 0] =$$

$$= \varphi_t\{p/Q, p/Q\}[0, 1] = 0. \quad \square$$

Remark: We can have, however,

$$\varphi_t\{p/Q' \vee \sim p/Q''\}[0] > 0$$

and, consequently,

$$\varphi_t\{p/Q' \vee \sim p/Q''\}[1] < 1. \qquad \square$$

Proposition 2.6 (*The principle of contradiction*):

$$\varphi_t\{p/Q \cdot \sim p/Q\}[0] = 1.$$

Proof: We have:

$$\varphi_t\{p/Q \cdot \sim p/Q\}[0] =$$

$$= \sum_{(k_1,k_2) \neq (1,1)} \varphi_t\{p/Q, \sim p/Q\}[k_1, k_2] =$$

$$= \sum_{(k_1,k_2) \neq (1,1)} \varphi_t\{p/Q, p/Q\}[k_1, \bar{k}_2] =$$

$$= \sum_{(k_1,k_2) \neq (1,0)} \varphi_t\{p/Q, p/Q\}[k_1, k_2] =$$

$$= \varphi_t\{p/Q, p/Q\}[1, 1] + \varphi_t\{p/Q, p/Q\}[0, 0] +$$

$$+ \varphi_t\{p/Q, p/Q\}[0, 1] = \varphi_t\{p/Q\}[1] + \varphi_t\{p/Q\}[0] = 1. \quad \square$$

Remark: In the relative logic, a sentence and its negation may be simultaneously true or simulaneously false under different circumstances, i.e.,

$$\varphi_t\{p/Q_1 \cdot \sim p/Q_2\}[1] > 0, \quad \text{or/and} \quad \varphi_t\{p/Q_1 \cdot \sim p/Q_2\}[0] > 0.$$

Proposition 2.7 (*Reductio ad absurdum principle*):

$$\varphi_t\{(p/Q \Rightarrow \sim p/Q) \Rightarrow \sim p/Q\}[1] = 1.$$

Proof: We have:

$$\varphi_t\{(p/Q \Rightarrow \sim p/Q) \Rightarrow \sim p/Q\}[0] =$$

$$= \varphi_t\{p/Q \Rightarrow \sim p/Q, \sim p/Q\}[1,0] =$$

$$= \varphi_t\{\sim p/Q \vee \sim p/Q, \sim p/Q\}[1,0] =$$

$$= \sum_{(k_1,k_2)\neq(0,0)} \varphi_t\{\sim p/Q, \sim p/Q, \sim p/Q\}[k_1, k_2, 0] =$$

$$= \varphi_t\{\sim p/Q, \sim p/Q, \sim p/Q\}[1,1,0] +$$

$$+ \varphi_t\{\sim p/Q, \sim p/Q, \sim p/Q\}[1,0,0] +$$

$$+ \varphi_t\{\sim p/Q, \sim p/Q, \sim p/Q\}[0,1,0] =$$

$$= \varphi_t\{p/Q, p/Q, p/Q\}[0,0,1] +$$

$$+ \varphi_t\{p/Q, p/Q, p/Q\}[0,1,1] +$$

$$+ \varphi_t\{p/Q, p/Q, p/Q\}[1,0,1] = 0. \quad \square$$

Remark: If a sentence implies its own negation, under the same circumstances, then it is false.

Proposition 2.8 (*The principle of syllogism - first variant*):

$$\varphi_t\{(p_2/Q_2 \Rightarrow p_3/Q_3) \Rightarrow ((p_1/Q_1 \Rightarrow p_2/Q_2) \Rightarrow (p_1/Q_1 \Rightarrow p_3/Q_3))\}[1] = 1.$$

Proof: We have:

$$\varphi_t\{(p_2/Q_2 \Rightarrow p_3/Q_3) \Rightarrow ((p_1/Q_1 \Rightarrow p_2/Q_2) \Rightarrow (p_1/Q_1 \Rightarrow p_3/Q_3))\}[0] =$$

$$= \varphi_t\{\sim (p_2/Q_2 \Rightarrow p_3/Q_3) \vee ((p_1/Q_1 \Rightarrow p_2/Q_2) \Rightarrow (p_1/Q_1 \Rightarrow p_3/Q_3))\}[0] =$$

$$= \varphi_t\{\sim (p_2/Q_2 \Rightarrow p_3/Q_3), ((p_1/Q_1 \Rightarrow p_2/Q_2) \Rightarrow (p_1/Q_1 \Rightarrow p_3/Q_3))\}[0,0] =$$

$$- \varphi_t\{p_2/Q_2 \rightarrow p_3/Q_3, \sim (p_1/Q_1 \Rightarrow p_2/Q_2) \vee (p_1/Q_1 \Rightarrow p_3/Q_3)\}[1,0] =$$

$$= \varphi_t\{\sim p_2/Q_2 \vee p_3/Q_3, \sim (p_1/Q_1 \Rightarrow p_2/Q_2), p_1/Q_1 \Rightarrow p_3/Q_3\}[1,0,0] =$$

$$= \sum_{(k,\ell)\neq(0,0)} \varphi_t\{\sim p_2/Q_2, p_3/Q_3, p_1/Q_1 \Rightarrow p_2/Q_2, p_1/Q_1 \Rightarrow p_3/Q_3\}[k, \ell, 1, 0] =$$

$$= \sum^{*} \varphi_t\{p_2/Q_2, p_3/Q_3, \sim p_1/Q_1, p_2/Q_2, \sim p_1/Q_1, p_3/Q_3\}[\bar{k}, \ell, m, n, 0, 0] =$$

$$= \sum^{**} \varphi_t\{p_2/Q_2, p_3/Q_3, p_1/Q_1, p_2/Q_2, p_1/Q_1, p_3/Q_3\}[k, \ell, m, n, 1, 0] = 0,$$

where:

$$\sum^{*} = \sum_{(k,\ell)\neq(0,0),(m,n)\neq(0,0)},$$

$$\sum^{**} = \sum_{(k,\ell)\neq(1,0),(m,n)\neq(1,0)}. \quad \square$$

Proposition 2.9 (*The principle of syllogism - second variant*):

$$\varphi_t\{(p_1/Q_1 \Rightarrow p_2/Q_2) \Rightarrow ((p_2/Q_2 \Rightarrow p_3/Q_3) \Rightarrow (p_1/Q_1 \Rightarrow p_3/Q_3))\}[1] = 1.$$

Proof: Similarly with the details given in the proof of Proposition 2.8, we have:

$$\varphi_t\{(p_1/Q_1 \Rightarrow p_2/Q_2) \Rightarrow ((p_2/Q_2 \Rightarrow p_3/Q_3) \Rightarrow (p_1/Q_1 \Rightarrow p_3/Q_3))\}[0] =$$

$$= \varphi_t\{p_1/Q_1 \Rightarrow p_2/Q_2, (p_2/Q_2 \Rightarrow p_3/Q_3) \Rightarrow (p_1/Q_1 \Rightarrow p_3/Q_3)\}[1, 0] =$$

$$= \sum_{(k,\ell)\neq(1,0)} \varphi_t\{p_1/Q_1, p_2/Q_2, p_2/Q_2 \Rightarrow p_3/Q_3, p_1/Q_1 \Rightarrow p_3/Q_3\}[k, \ell, 1, 0] =$$

$$= \sum^{*} \varphi_t\{p_1/Q_1, p_2/Q_2, p_2/Q_2, p_3/Q_3, p_1/Q_1, p_3/Q_3\}[k, \ell, m, n, 1, 0] = 0,$$

where:

$$\sum^{*} = \sum_{(k,\ell)\neq(1,0),(m,n)\neq(1,0)}. \quad \square$$

Remark: A syllogism is a deductive scheme of a formal argument consisting of a major and a minor premise and a conclusion, like: "Every virtue is laudable; kindness is a virtue; therefore kindness is laudable."

Proposition 2.10 (*Identity law*):

$$\varphi_t\{p/Q \Rightarrow p/Q\}[1] = 1.$$

Proof: We have:

$$\varphi_t\{p/Q \Rightarrow p/Q\}[0] = \varphi_t\{p/Q, p/Q\}[1, 0] = 0. \quad \square$$

Proposition 2.11 (*Transposition principle*):

$$\varphi_t\{(p_1/Q_1 \Rightarrow p_2/Q_2) \Rightarrow (\sim p_2/Q_2 \Rightarrow \sim p_1/Q_1)\}[1] = 1.$$

Proof: We have:

$$\varphi_t\{(p_1/Q_1 \Rightarrow p_2/Q_2) \Rightarrow (\sim p_2/Q_2 \Rightarrow \sim p_1/Q_1)\}[0] =$$

$$= \varphi_t\{p_1/Q_1 \Rightarrow p_2/Q_2, \sim p_2/Q_2 \Rightarrow \sim p_1/Q_1)\}[1,0] =$$

$$= \sum_{(k,\ell)\neq(1,0)} \varphi_t\{p_1/Q_1, p_2/Q_2, \sim p_2/Q_2, \sim p_1/Q_1)\}[k,\ell,1,0] =$$

$$= \sum_{(k,\ell)\neq(1,0)} \varphi_t\{p_1/Q_1, p_2/Q_2, p_2/Q_2, p_1/Q_1)\}[k,\ell,0,1] = 0. \quad \square$$

Proposition 2.12 (*Complementary reductio ad absurdum principle*):

$$\varphi_t\{(\sim p/Q \Rightarrow p/Q) \Rightarrow p/Q\}[1] = 1.$$

Proof: We have:

$$\varphi_t\{(\sim p/Q \Rightarrow p/Q) \Rightarrow p/Q\}[0] =$$

$$= \varphi_t\{\sim p/Q \Rightarrow p/Q, p/Q\}[1,0] =$$

$$= \sum_{(k,\ell)\neq(1,0)} \varphi_t\{\sim p/Q, p/Q, p/Q\}[k,\ell,0] =$$

$$= \sum_{(k,\ell)\neq(1,0)} \varphi_t\{p/Q, p/Q, p/Q\}[\bar{k},\ell,0] =$$

$$= \sum_{(k,\ell)\neq(0,0)} \varphi_t\{p/Q, p/Q, p/Q\}[k,\ell,0] = 0. \quad \square$$

Remark: In classic logic a sentence induced by its own falseness is true.

Proposition 2.13 (*Principle of simplification*):

$$\varphi_t\{(p_1/Q_1 \cdot p_2/Q_2) \Rightarrow p_1/Q_1\}[1] = 1.$$

Proof: We have:

$$\varphi_t\{(p_1/Q_1 \cdot p_2/Q_2) \Rightarrow p_1/Q_1\}[0] =$$

$$= \varphi_t\{p_1/Q_1 \cdot p_2/Q_2, p_1/Q_1\}[1, 0] =$$

$$= \varphi_t\{p_1/Q_1, p_2/Q_2, p_1/Q_1\}[1, 1, 0] = 0. \quad \square$$

Proposition 2.14 (*Export principle*):

$$\varphi_t\{(p_1/Q_1 \cdot p_2/Q_2 \Rightarrow p_3/Q_3) \Rightarrow (p_1/Q_1 \Rightarrow (p_2/Q_2 \Rightarrow p_3/Q_3))\}[1] = 1.$$

Proof: We have:

$$\varphi_t\{(p_1/Q_1 \cdot p_2/Q_2 \Rightarrow p_3/Q_3) \Rightarrow (p_1/Q_1 \Rightarrow (p_2/Q_2 \Rightarrow p_3/Q_3))\}[0] =$$

$$= \varphi_t\{p_1/Q_1 \cdot p_2/Q_2 \Rightarrow p_3/Q_3, p_1/Q_1 \Rightarrow (p_2/Q_2 \Rightarrow p_3/Q_3)\}[1, 0] =$$

$$= \sum_{(k,\ell)\neq(1,0)} \varphi_t\{p_1/Q_1 \cdot p_2/Q_2, p_3/Q_3, p_1/Q_1, p_2/Q_2 \Rightarrow p_3/Q_3\}[k, \ell, 1, 0] =$$

$$= \sum_{(k,\ell)\neq(1,0)} \varphi_t\{p_1/Q_1 \cdot p_2/Q_2, p_3/Q_3, p_1/Q_1, p_2/Q_2, p_3/Q_3\}[k, \ell, 1, 1, 0] = 0. \quad \square$$

Remark: In classic logic, Peano called this "the export principle" because p_1/Q_1 is taken out (exported) from the product $p_1/Q_1 \cdot p_2/Q_2$.

Proposition 2.15 (*Import principle*):

$$\varphi_t\{(p_1/Q_1 \Rightarrow (p_2/Q_2 \Rightarrow p_3/Q_3)) \Rightarrow (p_1/Q_1 \cdot p_2/Q_2 \Rightarrow p_3/Q_3)\}[1] = 1.$$

Proof: We have:

$$\varphi_t\{(p_1/Q_1 \Rightarrow (p_2/Q_2 \Rightarrow p_3/Q_3)) \Rightarrow (p_1/Q_1 \cdot p_2/Q_2 \Rightarrow p_3/Q_3)\}[0] =$$

$$= \varphi_t\{p_1/Q_1 \Rightarrow (p_2/Q_2 \Rightarrow p_3/Q_3), p_1/Q_1 \cdot p_2/Q_2 \Rightarrow p_3/Q_3\}[1,0] =$$

$$= \sum_{(k,\ell)\neq(1,0)} \varphi_t\{p_1/Q_1, p_2/Q_2 \Rightarrow p_3/Q_3, p_1/Q_1 \cdot p_2/Q_2, p_3/Q_3\}[k,\ell,1,0] =$$

$$= \sum_{(k,\ell)\neq(1,0)} \varphi_t\{p_1/Q_1, p_2/Q_2, p_3/Q_3, p_1/Q_1, p_2/Q_2, p_3/Q_3\}[1,k,\ell,1,1,0] = 0. \quad \square$$

Proposition 2.16 (*Principle of composition*):

$$\varphi_t\{((p_1/Q_1 \Rightarrow p_2/Q_2)\cdot(p_1/Q_1 \Rightarrow p_3/Q_3)) \Rightarrow (p_1/Q_1 \Rightarrow p_2/Q_2\cdot p_3/Q_3)\}[1] = 1.$$

Proof: We have:

$$\varphi_t\{((p_1/Q_1 \Rightarrow p_2/Q_2) \cdot (p_1/Q_1 \Rightarrow p_3/Q_3)) \Rightarrow (p_1/Q_1 \Rightarrow p_2/Q_2 \cdot p_3/Q_3)\}[0] =$$

$$= \varphi_t\{(p_1/Q_1 \Rightarrow p_2/Q_2) \cdot (p_1/Q_1 \Rightarrow p_3/Q_3), p_1/Q_1 \Rightarrow p_2/Q_2 \cdot p_3/Q_3\}[1,0] =$$

$$= \varphi_t\{p_1/Q_1 \Rightarrow p_2/Q_2, p_1/Q_1 \Rightarrow p_3/Q_3, p_1/Q_1, p_2/Q_2 \cdot p_3/Q_3\}[1,1,1,0] =$$

$$= \overset{*}{\sum} \varphi_t\{p_1/Q_1, p_2/Q_2, p_1/Q_1, p_3/Q_3, p_1/Q_1, p_2/Q_2, p_3/Q_3\}[k,\ell,m,n,1,r,s] = 0,$$

where:

$$\overset{*}{\sum} = \sum_{(k,\ell)\neq(1,0),(m,n)\neq(1,0),(r,s)\neq(1,1)} . \quad \square$$

Proposition 2.17 (*Factor principle*):

$$\varphi_t\{(p_1/Q_1 \Rightarrow p_2/Q_2) \Rightarrow (p_1/Q_1 \cdot p_3/Q_3 \Rightarrow p_2/Q_2 \cdot p_3/Q_3)\}[1] = 1.$$

Proof: We have:

$$\varphi_t\{(p_1/Q_1 \Rightarrow p_2/Q_2) \Rightarrow (p_1/Q_1 \cdot p_3/Q_3 \Rightarrow p_2/Q_2 \cdot p_3/Q_3)\}[0] =$$

$$= \varphi_t\{p_1/Q_1 \Rightarrow p_2/Q_2, p_1/Q_1 \cdot p_3/Q_3 \Rightarrow p_2/Q_2 \cdot p_3/Q_3\}[1,0] =$$

$$= \sum_{(k,\ell)\neq(1,0)} \varphi_t\{p_1/Q_1, p_2/Q_2, p_1/Q_1 \cdot p_3/Q_3, p_2/Q_2 \cdot p_3/Q_3\}[k,\ell,1,0] =$$

$$= \overset{*}{\sum} \varphi_t\{p_1/Q_1, p_2/Q_2, p_1/Q_1, p_3/Q_3, p_2/Q_2, p_3/Q_3\}[k,\ell,1,1,m,n] = 0,$$

where:

$$\overset{*}{\sum} = \sum_{(k,\ell)\neq(1,0),(m,n)\neq(1,1)} . \quad \square$$

Proposition 2.18 (*Equivalence*):

$$\varphi_t\{p_1/Q_1 \equiv p_2/Q_2\} = \varphi_t\{(p_1/Q_1 \Rightarrow p_2/Q_2) \cdot (p_2/Q_2 \Rightarrow p_1/Q_1)\}.$$

Proof. We have:

$$\varphi_t\{p_1/Q_1 \equiv p_2/Q_2\}[1] =$$
$$= \varphi_t\{p_1/Q_1, p_2/Q_2\}[1,1] + \varphi_t\{p_1/Q_1, p_2/Q_2\}[0,0].$$

On the other hand,

$$\varphi_t\{(p_1/Q_1 \Rightarrow p_2/Q_2) \cdot (p_2/Q_2 \Rightarrow p_1/Q_1)\}[1] =$$

$$= \varphi_t\{p_1/Q_1 \Rightarrow p_2/Q_2, p_2/Q_2 \Rightarrow p_1/Q_1\}[1,1] =$$

$$= \sum_{(k,\ell)\neq(1,0),(m,n)\neq(1,0)} \varphi_t\{p_1/Q_1, p_2/Q_2, p_2/Q_2, p_1/Q_1\}[k,\ell,m,n] =$$

$$= \varphi_t\{p_1/Q_1, p_2/Q_2\}[1,1] + \varphi_t\{p_1/Q_1, p_2/Q_2\}[0,0].$$

Therefore,

$$\varphi_t\{p_1/Q_1 \equiv p_2/Q_2\}[1] = \varphi_t\{(p_1/Q_1 \Rightarrow p_2/Q_2) \cdot (p_2/Q_2 \Rightarrow p_1/Q_1)\}[1],$$

which implies also:

$$\varphi_t\{p_1/Q_1 \equiv p_2/Q_2\}[0] = \varphi_t\{(p_1/Q_1 \Rightarrow p_2/Q_2) \cdot (p_2/Q_2 \Rightarrow p_1/Q_1)\}[0]. \quad \square$$

Proposition 2.19 (*Principle of transposition*):

$$\varphi_t\{(p_1/Q_1 \Rightarrow p_2/Q_2) \equiv (\sim p_2/Q_2 \Rightarrow \sim p_1/Q_1)\}[1] = 1.$$

Proof: We have:

$$\varphi_t\{(p_1/Q_1 \Rightarrow p_2/Q_2) \equiv (\sim p_2/Q_2 \Rightarrow \sim p_1/Q_1)\}[1] =$$

$$= \varphi_t\{p_1/Q_1 \Rightarrow p_2/Q_2, \sim p_2/Q_2 \Rightarrow \sim p_1/Q_1\}[1,1]+$$

$$+\varphi_t\{p_1/Q_1 \Rightarrow p_2/Q_2, \sim p_2/Q_2 \Rightarrow \sim p_1/Q_1\}[0,0] =$$

$$= \sum_{(k,\ell)\neq(1,0),(m,n)\neq(1,0)} \varphi_t\{p_1/Q_1, p_2/Q_2, \sim p_2/Q_2, \sim p_1/Q_1\}[k,\ell,m,n]+$$

$$+\varphi_t\{p_1/Q_1, p_2/Q_2, \sim p_2/Q_2, \sim p_1/Q_1\}[1,0,1,0] =$$

$$= \sum_{(k,\ell)\neq(1,0),(m,n)\neq(0,1)} \varphi_t\{p_1/Q_1, p_2/Q_2, p_2/Q_2, p_1/Q_1\}[k,\ell,m,n]+$$

$$+\varphi_t\{p_1/Q_1, p_2/Q_2, p_2/Q_2, p_1/Q_1\}[1,0,0,1] =$$

$$= \varphi_t\{p_1/Q_1, p_2/Q_2\}[1,1] + \varphi_t\{p_1/Q_1, p_2/Q_2\}[1,0]+$$

$$+\varphi_t\{p_1/Q_1, p_2/Q_2\}[0,1] + \varphi_t\{p_1/Q_1, p_2/Q_2\}[0,0] = 1. \quad \square$$

Proposition 2.20 (*Law of double negation*):

$$\varphi_t\{p/Q \equiv \sim (\sim p)/Q\}[1] = 1.$$

Proof: We have:
$$\varphi_t\{p/Q \equiv \sim (\sim p)/Q\}[1] =$$

$$= \varphi_t\{p/Q, \sim (\sim p)/Q\}[1,1] + \varphi_t\{p/Q, \sim (\sim p)/Q\}[0,0] =$$

$$= \varphi_t\{p/Q, \sim p/Q\}[1,0] + \varphi_t\{p/Q, \sim p/Q\}[0,1] =$$

$$= \varphi_t\{p/Q, p/Q\}[1,1] + \varphi_t\{p/Q, p/Q\}[0,0] =$$

$$= \varphi_t\{p/Q\}[1] + \varphi_t\{p/Q\}[0] = 1. \quad \square$$

Proposition 2.21 (*Law of tautology*):

$$\varphi_t\{p/Q \equiv (p/Q \cdot p/Q)\}[1] = 1,$$

$$\varphi_t\{p/Q \equiv (p/Q \vee p/Q)\}[1] = 1.$$

Proof: We have:

$$\varphi_t\{p/Q \equiv (p/Q \cdot p/Q)\}[1] =$$

$$= \varphi_t\{p/Q, p/Q \cdot p/Q\}[1,1] + \varphi_t\{p/Q, p/Q \cdot p/Q\}[0,0] =$$

$$= \varphi_t\{p/Q, p/Q, p/Q\}[1,1,1] + \sum_{(k,\ell)\neq(1,1)} \varphi_t\{p/Q, p/Q, p/Q\}[0,k,\ell] =$$

$$= \varphi_t\{p/Q\}[1] + \varphi_t\{p/Q\}[0] = 1.$$

Similarly,

$$\varphi_t\{p/Q \equiv (p/Q \vee p/Q)\}[1] =$$

$$= \varphi_t\{p/Q, p/Q \vee p/Q\}[1,1] + \varphi_t\{p/Q, p/Q \vee p/Q\}[0,0] =$$

$$= \sum_{(k,\ell)\neq(0,0)} \varphi_t\{p/Q, p/Q, p/Q\}[1,k,\ell] + \varphi_t\{p/Q, p/Q, p/Q\}[0,0,0] =$$

$$= \varphi_t\{p/Q\}[1] + \varphi_t\{p/Q\}[0] = 1. \quad \square$$

Proposition 2.22 (*First De Morgan Law*):

$$\varphi_t\{(\sim (p_1/Q_1 \cdot p_2/Q_2)) \equiv (\sim p_1/Q_1 \vee \sim p_2/Q_2)\}[1] = 1.$$

Proof: We have:

$$\varphi_t\{(\sim (p_1/Q_1 \cdot p_2/Q_2)) \equiv (\sim p_1/Q_1 \vee \sim p_2/Q_2)\}[1] =$$

$$= \varphi_t\{\sim (p_1/Q_1 \cdot p_2/Q_2), \sim p_1/Q_1 \vee \sim p_2/Q_2\}[1,1]+$$

$$+\varphi_t\{\sim (p_1/Q_1 \cdot p_2/Q_2), \sim p_1/Q_1 \vee \sim p_2/Q_2\}[0,0] =$$

$$= \varphi_t\{p_1/Q_1 \cdot p_2/Q_2, \sim p_1/Q_1 \vee \sim p_2/Q_2\}[0,1]+$$

$$+\varphi_t\{p_1/Q_1 \cdot p_2/Q_2, \sim p_1/Q_1 \vee \sim p_2/Q_2\}[1,0] =$$

$$= \sum_{(k,\ell)\neq(1,1),(m,n)\neq(0,0)} \varphi_t\{p_1/Q_1, p_2/Q_2, \sim p_1/Q_1, \sim p_2/Q_2\}[k,\ell,m,n]+$$

$$+\varphi_t\{p_1/Q_1, p_2/Q_2, \sim p_1/Q_1, \sim p_2/Q_2\}[1,1,0,0] =$$

$$= \sum_{(k,\ell)\neq(1,1),(m,n)\neq(1,1)} \varphi_t\{p_1/Q_1, p_2/Q_2, p_1/Q_1, p_2/Q_2\}[k,\ell,m,n]+$$

$$+\varphi_t\{p_1/Q_1, p_2/Q_2, p_1/Q_1, p_2/Q_2\}[1,1,1,1] =$$

$$= \varphi_t\{p_1/Q_1, p_2/Q_2\}[0,0] + \varphi_t\{p_1/Q_1, p_2/Q_2\}[1,0]+$$

$$+\varphi_t\{p_1/Q_1, p_2/Q_2\}[0,1] + \varphi_t\{p_1/Q_1, p_2/Q_2\}[1,1] = 1. \quad \square$$

Proposition 2.23 (*Second De Morgan Law*):

$$\varphi_t\{(\sim (p_1/Q_1 \vee p_2/Q_2)) \equiv (\sim p_1/Q_1 \cdot \sim p_2/Q_2)\}[1] = 1.$$

Proof: We have:

$$\varphi_t\{(\sim (p_1/Q_1 \vee p_2/Q_2)) \equiv (\sim p_1/Q_1 \cdot \sim p_2/Q_2)\}[1] =$$

$$= \varphi_t\{\sim (p_1/Q_1 \vee p_2/Q_2), \sim p_1/Q_1 \cdot \sim p_2/Q_2\}[1,1]+$$

$$+\varphi_t\{\sim (p_1/Q_1 \vee p_2/Q_2), \sim p_1/Q_1 \cdot \sim p_2/Q_2\}[0,0] =$$

$$= \varphi_t\{p_1/Q_1 \vee p_2/Q_2, \sim p_1/Q_1 \cdot \sim p_2/Q_2\}[0,1]+$$

$$+\varphi_t\{p_1/Q_1 \vee p_2/Q_2, \sim p_1/Q_1 \cdot \sim p_2/Q_2\}[1,0] =$$

$$= \varphi_t\{p_1/Q_1, p_2/Q_2, \sim p_1/Q_1, \sim p_2/Q_2\}[0,0,1,1]+$$

$$+ \sum_{(k,\ell)\neq(0,0),(m,n)\neq(1,1)} \varphi_t\{p_1/Q_1, p_2/Q_2, \sim p_1/Q_1, \sim p_2/Q_2\}[k,\ell,m,n] =$$

$$= \varphi_t\{p_1/Q_1, p_2/Q_2, p_1/Q_1, p_2/Q_2\}[0,0,0,0]+$$

$$+ \sum_{(k,\ell)\neq(0,0),(m,n)\neq(0,0)} \varphi_t\{p_1/Q_1, p_2/Q_2, p_1/Q_1, p_2/Q_2\}[k,\ell,m,n] =$$

$$= \varphi_t\{p_1/Q_1, p_2/Q_2\}[0,0] + \varphi_t\{p_1/Q_1, p_2/Q_2\}[1,1]+$$

$$+\varphi_t\{p_1/Q_1, p_2/Q_2\}[1,0] + \varphi_t\{p_1/Q_1, p_2/Q_2\}[0,1] = 1. \quad \square$$

Proposition 2.24 (*Implication*):

$$\varphi_t\{(p_1/Q_1 \Rightarrow p_2/Q_2) \equiv (p_1/Q_1 \equiv p_1/Q_1 \cdot p_2/Q_2)\}[1] = 1.$$

Proof: We have:

$$\varphi_t\{(p_1/Q_1 \Rightarrow p_2/Q_2) \equiv (p_1/Q_1 \equiv p_1/Q_1 \cdot p_2/Q_2)\}[1] =$$

$$= \varphi_t\{p_1/Q_1 \Rightarrow p_2/Q_2, p_1/Q_1 \equiv p_1/Q_1 \cdot p_2/Q_2\}[1,1]+$$

$$+\varphi_t\{p_1/Q_1 \Rightarrow p_2/Q_2, p_1/Q_1 \equiv p_1/Q_1 \cdot p_2/Q_2\}[0,0] =$$

$$= \sum_{(k,\ell)\neq(1,0)} \varphi_t\{p_1/Q_1, p_2/Q_2, p_1/Q_1, p_1/Q_1 \cdot p_2/Q_2\}[k,\ell,1,1]+$$

$$+ \sum_{(k,\ell)\neq(1,0)} \varphi_t\{p_1/Q_1, p_2/Q_2, p_1/Q_1, p_1/Q_1 \cdot p_2/Q_2\}[k,\ell,0,0]+$$

$$+\varphi_t\{p_1/Q_1, p_2/Q_2, p_1/Q_1, p_1/Q_1 \cdot p_2/Q_2\}[1,0,1,0]+$$

$$+\varphi_t\{p_1/Q_1, p_2/Q_2, p_1/Q_1, p_1/Q_1 \cdot p_2/Q_2\}[1,0,0,1] =$$

$$= \sum_{(k,\ell)\neq(1,0)} \varphi_t\{p_1/Q_1, p_2/Q_2, p_1/Q_1, p_1/Q_1, p_2/Q_2\}[k,\ell,1,1,1]+$$

$$+ \sum_{(k,\ell)\neq(1,0),(m,n)\neq(1,1)} \varphi_t\{p_1/Q_1, p_2/Q_2, p_1/Q_1, p_1/Q_1, p_2/Q_2\}[k,\ell,0,m,n]+$$

$$+ \sum_{(m,n)\neq(1,1)} \varphi_t\{p_1/Q_1, p_2/Q_2, p_1/Q_1, p_1/Q_1, p_2/Q_2\}[1,0,1,m,n]+$$

$$+\varphi_t\{p_1/Q_1, p_2/Q_2, p_1/Q_1, p_1/Q_1, p_2/Q_2\}[1,0,0,1,1] =$$

$$= \varphi_t\{p_1/Q_1, p_2/Q_2\}[1,1] + \varphi_t\{p_1/Q_1, p_2/Q_2\}[1,0]+$$

$$+\varphi_t\{p_1/Q_1, p_2/Q_2\}[0,1] + \varphi_t\{p_1/Q_1, p_2/Q_2\}[0,0] = 1. \quad \square$$

Remark: This property shows how an implication may be reduced to an equivalence.

Proposition 2.25:

$$\varphi_t\{\sim ((p_1/Q_1 \equiv p_2/Q_2) \cdot (p_1/Q_1 \equiv\sim p_2/Q_2))\}[1] = 1.$$

Proof: We have:

$$\varphi_t\{\sim ((p_1/Q_1 \equiv p_2/Q_2) \cdot (p_1/Q_1 \equiv\sim p_2/Q_2))\}[0] =$$

$$= \varphi_t\{(p_1/Q_1 \equiv p_2/Q_2) \cdot (p_1/Q_1 \equiv\sim p_2/Q_2)\}[1] =$$

$$= \varphi_t\{p_1/Q_1 \equiv p_2/Q_2, p_1/Q_1 \equiv\sim p_2/Q_2\}[1,1] =$$

$$= \varphi_t\{p_1/Q_1, p_2/Q_2, p_1/Q_1, \sim p_2/Q_2\}[1,1,1,1]+$$

$$+\varphi_t\{p_1/Q_1, p_2/Q_2, p_1/Q_1, \sim p_2/Q_2\}[1,1,0,0]+$$

$$+\varphi_t\{p_1/Q_1, p_2/Q_2, p_1/Q_1, \sim p_2/Q_2\}[0,0,1,1]+$$
$$+\varphi_t\{p_1/Q_1, p_2/Q_2, p_1/Q_1, \sim p_2/Q_2\}[0,0,0,0]=$$
$$= \varphi_t\{p_1/Q_1, p_2/Q_2, p_1/Q_1, p_2/Q_2\}[1,1,1,0]+$$
$$+\varphi_t\{p_1/Q_1, p_2/Q_2, p_1/Q_1, p_2/Q_2\}[1,1,0,1]+$$
$$+\varphi_t\{p_1/Q_1, p_2/Q_2, p_1/Q_1, p_2/Q_2\}[0,0,1,0]+$$
$$+\varphi_t\{p_1/Q_1, p_2/Q_2, p_1/Q_1, p_2/Q_2\}[0,0,0,1]=0. \quad \square$$

Remarks: 1) It is not true that a sentence is equivalent both to another sentence and to its negation, under the same circumstances.

2) We may also write:

$$\varphi_t\{(p_1/Q_1 \equiv p_2/Q_2) \cdot (p_1/Q_1 \equiv \sim p_2/Q_2)\}[0] = 1,$$

showing that it is certain that a sentence cannot be equivalent both to another sentence and to the negation of it, under the same circumstances.

3) We may have, however,

$$\varphi_t\{(p_1/Q_1 \equiv p_2/Q_2') \cdot (p_1/Q_1 \equiv \sim p_2/Q_2'')\}[1] > 0,$$

if the circumstances Q_2' and Q_2'' are not identical.

Proposition 2.26:

$$\varphi_t\{(p_1/Q_1 \vee p_2/Q_2) \equiv ((p_1/Q_1 \Rightarrow p_2/Q_2) \Rightarrow p_2/Q_2)\}[1] = 1.$$

Proof: We have:

$$\varphi_t\{(p_1/Q_1 \vee p_2/Q_2) \equiv ((p_1/Q_1 \Rightarrow p_2/Q_2) \Rightarrow p_2/Q_2)\}[0] =$$

$$= \varphi_t\{p_1/Q_1 \vee p_2/Q_2, (p_1/Q_1 \Rightarrow p_2/Q_2) \Rightarrow p_2/Q_2\}[1,0]+$$

$$+\varphi_t\{p_1/Q_1 \vee p_2/Q_2, (p_1/Q_1 \Rightarrow p_2/Q_2) \Rightarrow p_2/Q_2\}[0,1] =$$

$$= \sum_{(k,\ell)\neq(0,0)} \varphi_t\{p_1/Q_1, p_2/Q_2, p_1/Q_1 \Rightarrow p_2/Q_2, p_2/Q_2\}[k,\ell,1,0]+$$

$$+ \sum_{(k,\ell)\neq(1,0)} \varphi_t\{p_1/Q_1, p_2/Q_2, p_1/Q_1 \Rightarrow p_2/Q_2, p_2/Q_2\}[0,0,k,\ell] =$$

$$= \varphi_t\{p_1/Q_1, p_2/Q_2, p_1/Q_1 \Rightarrow p_2/Q_2, p_2/Q_2\}[1,0,1,0]+$$

$$+\varphi_t\{p_1/Q_1, p_2/Q_2, p_1/Q_1 \Rightarrow p_2/Q_2, p_2/Q_2\}[0,0,0,0] =$$

$$= \sum_{(k,\ell)\neq(1,0)} \varphi_t\{p_1/Q_1, p_2/Q_2, p_1/Q_1, p_2/Q_2, p_2/Q_2\}[1,0,k,\ell,0]+$$

$$+\varphi_t\{p_1/Q_1, p_2/Q_2, p_1/Q_1, p_2/Q_2, p_2/Q_2\}[0,0,1,0,0] = 0. \quad \Box$$

Remark: This property shows that it is certain that the disjunction is equivalent to two implications.

Proposition 2.27:

$$\varphi_t\{\sim p_1/Q_1 \Rightarrow (p_1/Q_1 \Rightarrow p_2/Q_2)\}[1] = 1.$$

Proof: We have:

$$\varphi_t\{\sim p_1/Q_1 \Rightarrow (p_1/Q_1 \Rightarrow p_2/Q_2)\}[0] =$$

$$= \varphi_t\{\sim p_1/Q_1, p_1/Q_1 \Rightarrow p_2/Q_2\}[1,0] =$$

$$= \varphi_t\{p_1/Q_1, p_1/Q_1, p_2/Q_2\}[0,1,0] = 0. \quad \Box$$

2.8. Measures of uncertainty and interdependence

Given the logic probability (credibility) distribution

$$\varphi_t\{p/Q\},$$

the *entropy* of sentence p, under circumstances Q, at time t is the nonnegative number:

$$H_t(p/Q) = H_t(\varphi_t\{p/Q\}) = -\sum_k \varphi_t\{p/Q\}[k] \log \varphi_t\{p/Q\}[k],$$

where the base of the log is either 2, or 10, or the transcendental number e, depending on the respective application we are dealing with. The entropy H_t measures the amount of uncertainty contained by the logic probability (credibility) distribution $\varphi_t\{p/Q\}$. Its basic properties do not depend on the base of the logarithm, assumed to be larger than 1. In any application, however, once the base of the logarithm is chosen, it will remain the same during the all subsequent computations. In what follows, we assume that the base of the logarithm is e, which corresponds to the so called natural logarithm, sometimes denoted by ln.

Given the joint logic probability (credibility) distribution $\varphi_t\{p_1/Q_1, p_2/Q_2\}$, the *joint entropy* of sentence p_1, under circumstances Q_1, and sentence p_2, under circumstances Q_2, at time t, is the nonnegative number:

$$H_t(p_1/Q_1, p_2/Q_2) =$$

$$= -\sum_{k,\ell} \varphi_t\{p_1/Q_1, p_2/Q_2\}[k, \ell] \log \varphi_t\{p_1/Q_1, p_2/Q_2\}[k, \ell],$$

where the sum is taken with respect to all possible truth (logical) values k of sentence p_1, under circumstances Q_1, at time t, and all possible truth (logical) values ℓ of sentence p_2, under circumstances Q_2, at time t. It measures the amount of uncertainty on the truth (logic) values of sentence p_1, under circumstances Q_1, and the truth (logic) values of sentence p_2, under circumstances Q_2, at time t.

Given the conditional logic probability (credibility) distributions:

$$\varphi_t\{p_1/Q_1 \mid p_2/Q_2\}[k \mid \ell] = \frac{\varphi_t\{p_1/Q_1, p_2/Q_2\}[k, \ell]}{\varphi_t\{p_2/Q_2\}[\ell]},$$

where the marginal probability (credibility) distribution:

$$\varphi_t\{p_2/Q_2\}[\ell] = \sum_k \varphi_t\{p_1/Q_1, p_2, Q_2\}[k, \ell],$$

the *conditional entropy* of sentence p_1, under circumstances Q_1, given sentence p_2, under circumstances Q_2, at time t, is the nonnegative number:

$$H_t(p_1/Q_1 \mid p_2/Q_2) =$$

$$= -\sum_{k,\ell} \varphi_t\{p_1/Q_1, p_2/Q_2\}[k, \ell] \log \varphi_t\{p_1/Q_1 \mid p_2/Q_2\}[k \mid \ell],$$

where the sum is taken with respect to all possible truth (logic) values k of sentence p_1, under circumstances Q_1, and the values ℓ of sentence p_2, under circumstances Q_2. It measures the amount of uncertainty on the truth (logic) values of sentence p_1, under circumstances Q_1, given (i.e., conditioned by) the truth (logic) values of sentence p_2, under circumstances Q_2, at time t.

Given the joint logic probability (credibility) distribution:

$$\varphi_t\{p_1/Q_1, p_2/Q_2\}[k, \ell],$$

the amount of *global interdependence* between the possible logical values of sentence p_1, under circumstances Q_1, at time t, and the possible logical values of sentence p_2, under circumstances Q_2, at time t, is the nonnegative number:

$$W_t(p_1/Q_1, p_2/Q_2) =$$
$$= H_t(p_1/Q_1) + H_t(p_2/Q_2) - H_t(p_1/Q_1, p_2/Q_2) =$$
$$= \sum_{k,\ell} \varphi_t\{p_1/Q_1, p_2/Q_2\}[k, \ell] \log \frac{\varphi_t\{p_1/Q_1, p_2/Q_2\}[k, \ell]}{\varphi_t\{p_1/Q_1\}[k]\, \varphi_t\{p_2/Q_2\}[\ell]},$$

where:

$$\varphi_t\{p_1/Q_1\}[k], \quad \varphi_t\{p_2/Q_2\}[\ell]$$

are the marginal logic probability (credibility) distributions of the joint distribution $\varphi_t\{p_1/Q_1, p_2/Q_2\}[k, \ell]$.

With the above notation, the χ^2-measure of *deviation from independence* of the possible logical values of sentence p_1, under circumstances Q_1, at time t, and the possible logical values of sentence p_2, under circumstances Q_2, at time t, is:

$$\chi_t^2(p_1/Q_1, p_2/Q_2) =$$
$$= \sum_{k,\ell} \frac{(\varphi_t\{p_1/Q_1, p_2/Q_2\}[k, \ell] - \varphi_t\{p_1/Q_1\}[k]\, \varphi_t\{p_2/Q_2\}[\ell])^2}{\varphi_t\{p_1/Q_1\}[k]\, \varphi_t\{p_2/Q_2\}[\ell]}.$$

Obviously, $\chi_t^2(p_1/Q_1, p_2/Q_2)$ is a nonnegative number and it is equal to zero if and only if the possible logical values of sentence p_1, under circumstances Q_1, and the possible logical values of sentence p_2, under circumstances Q_2, are independent at time t. Being a quadratic function, the χ_t^2-indicator, also known in statistics as Karl Pearson's indicator (Pearson, (1900)), it is very convenient to use it in optimization with mean values as constraints, because its partial derivatives with respect to the unknown components of a probability distribution are linear functions.

Chapter 3. MULTI-VALUED RELATIVE LOGIC

3.1. Łukasiewicz's three-valued logic

The classic logic assumes that there are only two values for sentences, i.e., true (1) and false (0), and is based on the principle of excluded third. Aristotle pleaded in favour of this principle but showed that there are difficulties when it is applied to future events. Thus, the sentence "It is possible that I will be in Warsaw on 21st December," is not true or false at the moment we state it; therefore it is possible but not necessary. In order to cope with such situations, Łukasiewicz (1930, 1957) introduced a third logical value, "possible", denoted by 1/2, founding a logic with three values. The negation $\sim p$ (non-p) and the implication $p_1 \Longrightarrow p_2$ (p_1 implies p_2), denoted by Łukasiewicz with the symbols Np and Cp_1p_2, respectively, are defined by the following tableaus:

p	0	1/2	1
Np	1	1/2	0

	Cp_1p_2	p_2: 0	1/2	1
p_1	0	1	1	1
	1/2	1/2	1	1
	1	0	1/2	1

Let un notice that for Łukasiewicz, writing p means both the sentence p and the statement "the sentence p is true," which sometimes could be confusing. The operators N and C may be used for defining the disjunction A, conjunction K, and equivalence E of two sentences, as follows:

$$Ap_1p_2 = CCp_1p_2p_2, \qquad (p_1 \text{ or } p_2),$$

$$Kp_1p_2 = NANp_1Np_2, \qquad (p_1 \text{ and } p_2),$$

$$Ep_1p_2 = KCp_1p_2Cp_2p_1, \qquad (p_1 \text{ and } p_2 \text{ are equivalent}).$$

Therefore, the logical values of A, K, and E are:

	Ap_1p_2	0	p_2 1/2	1
	0	0	1/2	1
p_1	1/2	1/2	1/2	1
	1	1	1	1

	Kp_1p_2	0	p_2 1/2	1
	0	0	0	0
p_1	1/2	0	1/2	1/2
	1	0	1/2	1

	Ep_1p_2	0	p_2 1/2	1
	0	1	1/2	0
p_1	1/2	1/2	1	1/2
	1	0	1/2	1

Except the possibility M, another modality may be introduced, namely, the *doubt D* of sentence p, defined by

$$Dp = EpNp.$$

The following tableau summarizes the logical values of different modalities in Lukasiewicz's three-dimensional logic:

p	Np	Mp	NMp	MNp	$NMNp$	Dp	NDp
0	1	0	1	1	0	0	1
1/2	1/2	1	0	1	0	1	0
1	0	1	0	0	1	0	1

Lukasiewicz's pupil, Alfred Tarski, defined the *possibility* of sentence p by using the negation and implication, i.e.,

$$Mp = CNpp,$$

meaning "if non-p, then p," implying that the statement Mp is false only when p is false. Its logical values are:

$$M0 = 0, \quad M\frac{1}{2} = 1, \quad M1 = 1,$$

showing that the possibility of the falsehood is false, the possibility of the possible is true, and the possibility of the truth is true. The operators N and M may be combined. Thus, NMp means "p is not possible," MNp "non-p is posible," $NMNp$ "non-p is not possible," or "it is not possible that p is false," or "p is necessarily true," or "p is necessary."

According to Łukasiewicz *necessity* may be defined using only the negation N and implication C, i.e.,

$$NMNp = NCpNp,$$

saying that "p is necessary" is the same as "it is not true that if p then non-p." A sentence is necessary only when it is false that it implies its own negation.

The axiomatic method starts from a set of postulates and rules from which theorems are obtained as consequences. Another approach is to look at theorems as being those truth functions (also called tautologies) that take on the value 1 (true) for all logical values of the variables. The proof of a theorem is provided by the truth matrix of the respective truth function, which is a tableau containing the values of respective truth function for each system of possible logical values of its variables. This is the reason why this second method is called the matrix (or matricial) method. Łukasiewicz consistently used the matrix method in proving the theorems of his three-valued logic. Thus, for instance, here are two theorems in Łukasiewicz's three-valued logic and their proofs using the matricial method:

Proposition 3.1: *The sentence:*

$$CpMp \quad or, \; equivalently, \quad CpCNpp$$

is a theorem in the three-valued logic.

Proof. Using the definition of the implication C from the tableau given before, we have the following truth matrix:

p	Np	$CNpp$	$CpCNpp$
0	1	0	1
1/2	1/2	1	1
1	0	1	1

The last column of the tableau shows that $CpMp$ is true whatever the truth value of sentence p is.   □

Proposition 3.2: *The sentence*:

$$CCp_1p_2CNp_2Np_1$$

is a theorem, stating that if p_1 implies p_2, then non-p_2 *implies* non-p_1.

Proof: We have the following truth matrix:

p_1	p_2	Cp_1p_2	Np_2	Np_1	CNp_2Np_1	$CCp_1p_2CNp_2Np_1$
0	0	1	1	1	1	1
1/2	0	1/2	1	1/2	1/2	1
1	0	0	1	0	0	1
0	1/2	1	1/2	1	1	1
1/2	1/2	1	1/2	1/2	1	1
1	1/2	1/2	1/2	0	1/2	1
0	1	1	0	1	1	1
1/2	1	1	0	1/2	1	1
1	1	1	0	0	1	1

The last column of the matrix contains only the value 1 (true) whatever the logical values of p_1 and p_2 are.   □

Similarly, we can prove the following theorems:

$CpCNpp$;

$CpMp$;

$CNMpNp$,

("if p is not possible then non-p");

$CNpCNpNMp$,

("if non-p, then non-p implies the impossibility of p");

$CpCpNMNp$,

("if p, then p implies the necessity of p");

Cpp,

(the identity law);

$CCp_1p_2CCp_2p_3Cp_1p_3$,

(first variant of the syllogism);

$CCp_2p_3CCp_1p_2Cp_1p_3$,
(second variant of the syllogism);
 $CNp_1Cp_1p_2$,
("a false sentence implies anything");
 $EpNNp$,
(the principle of double negation);
 $CDpMp$,
("doubtful implies possible");
 $CDpMNp$,
("doubtful implies possible non");
 $EDpDNp$,
("the doubt of p is equivalent to the doubt of non-p");
 $ENDpNDNp$,
("the lack of doubt on p is equivalent to the lack of doubt on Np");
 $AANMNpDpNMp$,
("any sentence p is either necessary, or doubtful, or impossible").

Let us notice, however, that the principle of excluded third, namely,

$$ApNp$$

is not a tautology (theorem) in the three-valued logic, because it takes the value $1/2$ when the logical value of p is $1/2$. This "tertium non datur" principle is replaced by "quartum non datur" principle, according to which any sentence p may have one of three logical values, 0, $1/2$, or 1.

In classic logic, the implication:

$$(\sim p \supset p) \supset p$$

is a special case of *reductio ad absurdum*: "If a sentence p is the consequence of its own falsehood then it is true." In Łukasiewicz's formalism, its analog, i.e.,

$$CCNppp,$$

is not true for the logical value $1/2$ of p and, therefore, it is not a theorem.

The principle of contradiction:

$$NKpNp$$

is not a tautology either; for the value $1/2$ of sentence p, the above truth function takes on the logical value $1/2$ as well.

3.2. Multi-valued relative logic

Let V be the set of possible logical values. In the m-valued logic we generally take V to be:

$$V = \{0, \frac{1}{m-1}, \frac{2}{m-1}, \ldots, \frac{m-2}{m-1}, 1\},$$

where 0 is "false", 1 is "true", and the other elements correspond to the intermediate truth values. In a continuously infinite-valued logic $V = [0, 1]$, where again 0 is "false", 1 is "true", and the intermediary truth values are arbitrary real numbers from the open interval $(0, 1)$.

If t_1 and t_2 are two arbitrary instants of time, then:

$$\varphi_{t_1, t_2}\{(p_{11}/Q_{11}, p_{21}/Q_{21}), (p_{12}/Q_{12}, p_{22}/Q_{22})\}[(k_{11}, k_{21}), (k_{12}, k_{22})]$$

is a number from the interval $[0, 1]$, representing the probability that, at time t_1, the sentence p_{11}, under circumstances Q_{11}, and the sentence p_{21}, under circumstances Q_{21}, take on the logical values k_{11}, k_{21}, respectively, and, at time t_2, the sentence p_{12}, under circumstances Q_{12}, and the sentence p_{22}, under circumstances Q_{22}, take on the logical values k_{12}, k_{22}, respectively. φ_{t_1, t_2} is the logical probability at times t_1, t_2. The following sum:

$$\sum_{k_{11}, k_{21}, k_{12}, k_{22}} \varphi_{t_1, t_2}\{(p_{11}/Q_{11}, p_{21}/Q_{21}), (p_{12}/Q_{12}, p_{22}/Q_{22})\}[(k_{11}, k_{21}), (k_{12}, k_{22})]$$

is equal to 1. The conditional logical probability (credibility) distribution at time t of the truth (or logical) value k of sentence p_1, under circumstances Q_1, given the truth (logical) value ℓ of sentence p_2, under circumstances Q_2, is defined by:

$$\varphi_t\{p_1/Q_1 \mid p_2/Q_2\}[k \mid \ell] = \frac{\varphi_t\{p_1/Q_1, p_2/Q_2\}[k, \ell]}{\varphi_t\{p_2/Q_2\}[\ell]},$$

for all possible truth (logical) values k of sentence p_1, assuming that the denominator is different from zero. For each truth (logical) value ℓ of sentence p_2, under circumstances Q_2,

$$\varphi_t\{p_1/Q_1 \mid p_2/Q_2\}[k \mid \ell], \quad k \in V,$$

is a probability (credibility) distribution on the truth (logic) values of sentence p_1, under circumstances Q_1.

Similarly, the conditional logical probability (credibility) distribution at time t_1 of the truth (or logical) value k of sentence p_1, under circumstances Q_1, given the truth (logical) value ℓ of sentence p_2, under circumstances Q_2, at time t_2, is defined by:

$$\varphi_{t_1|t_2}\{p_1/Q_1 \mid p_2/Q_2\}[k \mid \ell] = \frac{\varphi_{t_1,t_2}\{p_1/Q_1, p_2/Q_2\}[k, \ell]}{\varphi_{t_2}\{p_2/Q_2\}[\ell]},$$

for all possible truth (logical) values k of sentence p_1, assuming that the denominator is different from zero. For each truth (logical) value ℓ of sentence p_2, under circumstances Q_2, at time t_2,

$$\varphi_{t_1|t_2}\{p_1/Q_1 \mid p_2/Q_2\}[k \mid \ell]$$

is a probability (credibility) distribution on the truth (logic) values of sentence p_1, under circumstances Q_1, at time t_1.

The truth (logical) value k of sentence P is *certain, highly probable, lowly probable,* or *impossible* at time t if

$$\varphi_t\{P/Q\}[k] = 1, \quad 0.5 \leq \varphi_t\{P/Q\}[k] < 1,$$

$$0 < \varphi_t\{P/Q\}[k] < 0.5, \quad \varphi_t\{P/Q\}[k] = 0,$$

respectively. We write

$$\varphi\{P/Q\}, \qquad \varphi_t\{P\}, \qquad \varphi\{P\},$$

if time t, or circumstances Q, or both time t and circumstances Q are irrelevant or obviously understood, respectively.

In order to be more specific, this chapter will focus mainly on the three-valued relative logic, where $V = \{0, 1/2, 1\}$ (with $0 =$ false, $1/2 =$ possible, $1 =$ true), and on the four-valued relative logic, where $V = \{0, 1/3, 2/3, 1\}$ (with $0 =$ false, $1/3 =$ lowly probable, $2/3 =$ highly probable, $1 =$ true). With minor exceptions, the generalization to relative logics with more than four logical values is straightforward.

3.3. Operations

Negation $\sim$ *(non)*.

(a) In the three-valued relative logic:

$$\varphi_t\{\sim p/Q\}[0] = \varphi_t\{p/Q\}[1],$$

$$\varphi_t\{\sim p/Q\}[1/2] = \varphi_t\{p/Q\}[1/2],$$
$$\varphi_t\{\sim p/Q\}[1] = \varphi_t\{p/Q\}[0].$$

(b) In the four-valued relative logic:

$$\varphi_t\{\sim p/Q\}[0] = \varphi_t\{p/Q\}[1],$$

$$\varphi_t\{\sim p/Q\}[1/3] = \varphi_t\{p/Q\}[2/3],$$
$$\varphi_t\{\sim p/Q\}[2/3] = \varphi_t\{p/Q\}[1/3],$$
$$\varphi_t\{\sim p/Q\}[1] = \varphi_t\{p/Q\}[0].$$

(c) In the m-valued logic:

$$\varphi_t\{\sim p/Q\}[k] = \varphi_t\{(\sim p)/Q\}[k] = \varphi_t\{p/Q\}[1-k].$$

Remark: The negation just defined is the so called "diametral negation". Reichenbach (1949) also proposed two other negations, namely the "cyclic negation" and the "complete negation" which, in the context of the three-valued relative logic, would be

$$\varphi_t\{-p/Q\}[0] = \varphi_t\{p/Q\}[1/2],$$

$$\varphi_t\{-p/Q\}[1/2] = \varphi_t\{p/Q\}[1],$$
$$\varphi_t\{-p/Q\}[1] = \varphi_t\{p/Q\}[0],$$

and

$$\varphi_t\{\bar{p}/Q\}[1] = \varphi_t\{p/Q\}[1/2] + \varphi_t\{p/Q\}[0],$$
$$\varphi_t\{\bar{p}/Q\}[1/2] = \varphi_t\{p/Q\}[1],$$

respectively, where 0 is "false", 1 is "true", and $1/2$ is "undetermined".

Disjunction (logical sum) $\vee$ (*or*).

(a) In the three-valued logic:

$$\varphi_t\{p_1/Q_1 \vee p_2/Q_2\}[0] = \varphi_t\{p_1/Q_1, p_2/Q_2\}[0,0],$$

$$\varphi_t\{p_1/Q_1 \vee p_2/Q_2\}[1/2] = \varphi_t\{p_1/Q_1, p_2/Q_2\}[1/2, 1/2]+$$
$$+\varphi_t\{p_1/Q_1, p_2/Q_2\}[0, 1/2] + \varphi_t\{p_1/Q_1, p_2/Q_2\}[1/2, 0],$$
$$\varphi_t\{p_1/Q_1 \vee p_2/Q_2\}[1] = \varphi_t\{p_1/Q_1, p_2/Q_2\}[0, 1]+$$
$$+\varphi_t\{p_1/Q_1, p_2/Q_2\}[1, 0] + \varphi_t\{p_1/Q_1, p_2/Q_2\}[1/2, 1]+$$
$$+\varphi_t\{p_1/Q_1, p_2/Q_2\}[1, 1/2] + \varphi_t\{p_1/Q_1, p_2/Q_2\}[1, 1].$$

(b) In the four-valued logic:

$$\varphi_t\{p_1/Q_1 \vee p_2/Q_2\}[0] = \varphi_t\{p_1/Q_1, p_2/Q_2\}[0,0],$$

$$\varphi_t\{p_1/Q_1 \vee p_2/Q_2\}[1/3] = \varphi_t\{p_1/Q_1, p_2/Q_2\}[0, 1/3]+$$
$$+\varphi_t\{p_1/Q_1, p_2/Q_2\}[1/3, 0] + \varphi_t\{p_1/Q_1, p_2/Q_2\}[1/3, 1/3],$$
$$\varphi_t\{p_1/Q_1 \vee p_2/Q_2\}[2/3] = \varphi_t\{p_1/Q_1, p_2/Q_2\}[0, 2/3]+$$
$$+\varphi_t\{p_1/Q_1, p_2/Q_2\}[2/3, 0] + \varphi_t\{p_1/Q_1, p_2/Q_2\}[1/3, 2/3]+$$
$$+\varphi_t\{p_1/Q_1, p_2/Q_2\}[2/3, 1/3] + \varphi_t\{p_1/Q_1, p_2/Q_2\}[2/3, 2/3],$$
$$\varphi_t\{p_1/Q_1 \vee p_2/Q_2\}[1] = \varphi_t\{p_1/Q_1, p_2/Q_2\}[0, 1]+$$
$$+\varphi_t\{p_1/Q_1, p_2/Q_2\}[1, 0] + \varphi_t\{p_1/Q_1, p_2/Q_2\}[1/3, 1]+$$
$$+\varphi_t\{p_1/Q_1, p_2/Q_2\}[1, 1/3] + \varphi_t\{p_1/Q_1, p_2/Q_2\}[2/3, 1]+$$
$$+\varphi_t\{p_1/Q_1, p_2/Q_2\}[1, 2/3] + \varphi_t\{p_1/Q_1, p_2/Q_2\}[1, 1].$$

(c) In the m-valued logic:

$$\varphi_t\{p_1/Q_1 \vee p_2/Q_2\}[k] =$$

$$= \sum_{\{(k_1,k_2); \max\{k_1,k_2\}=k\}} \varphi_t\{p_1/Q_1, p_2/Q_2\}[k_1, k_2].$$

Conjunction (logical product) $\cdot$ (*and*).

(a) In the three-valued logic:

$$\varphi_t\{p_1/Q_1 \cdot p_2/Q_2\}[1] = \varphi_t\{p_1/Q_1, p_2/Q_2\}[1,1],$$

$$\varphi_t\{p_1/Q_1 \cdot p_2/Q_2\}[1/2] = \varphi_t\{p_1/Q_1, p_2/Q_2\}[1/2, 1/2]+$$
$$+\varphi_t\{p_1/Q_1, p_2/Q_2\}[1/2, 1] + \varphi_t\{p_1/Q_1, p_2/Q_2\}[1, 1/2],$$
$$\varphi_t\{p_1/Q_1 \cdot p_2/Q_2\}[0] = \varphi_t\{p_1/Q_1, p_2/Q_2\}[0, 1]+$$
$$+\varphi_t\{p_1/Q_1, p_2/Q_2\}[1, 0] + \varphi_t\{p_1/Q_1, p_2/Q_2\}[0, 1/2]+$$
$$+\varphi_t\{p_1/Q_1, p_2/Q_2\}[1/2, 0] + \varphi_t\{p_1/Q_1, p_2/Q_2\}[0, 0].$$

(b) In the four-valued logic:

$$\varphi_t\{p_1/Q_1 \cdot p_2/Q_2\}[1] = \varphi_t\{p_1/Q_1, p_2/Q_2\}[1,1],$$

$$\varphi_t\{p_1/Q_1 \cdot p_2/Q_2\}[2/3] = \varphi_t\{p_1/Q_1, p_2/Q_2\}[2/3, 1]+$$
$$+\varphi_t\{p_1/Q_1, p_2/Q_2\}[1, 2/3] + \varphi_t\{p_1/Q_1, p_2/Q_2\}[2/3, 2/3],$$
$$\varphi_t\{p_1/Q_1 \cdot p_2/Q_2\}[1/3] = \varphi_t\{p_1/Q_1, p_2/Q_2\}[1/3, 1]+$$
$$+\varphi_t\{p_1/Q_1, p_2/Q_2\}[1, 1/3] + \varphi_t\{p_1/Q_1, p_2/Q_2\}[1/3, 2/3]+$$
$$+\varphi_t\{p_1/Q_1, p_2/Q_2\}[2/3, 1/3] + \varphi_t\{p_1/Q_1, p_2/Q_2\}[1/3, 1/3],$$
$$\varphi_t\{p_1/Q_1 \cdot p_2/Q_2\}[0] = \varphi_t\{p_1/Q_1, p_2/Q_2\}[0, 1]+$$
$$+\varphi_t\{p_1/Q_1, p_2/Q_2\}[1, 0] + \varphi_t\{p_1/Q_1, p_2/Q_2\}[0, 2/3]+$$
$$+\varphi_t\{p_1/Q_1, p_2/Q_2\}[2/3, 0] + \varphi_t\{p_1/Q_1, p_2/Q_2\}[0, 1/3]+$$
$$+\varphi_t\{p_1/Q_1, p_2/Q_2\}[1/3, 0] + \varphi_t\{p_1/Q_1, p_2/Q_2\}[0, 0].$$

(c) In the *m*-valued logic:

$$\varphi_t\{p_1/Q_1 \cdot p_2/Q_2\}[k] =$$

$$= \sum_{\{(k_1,k_2); \min\{k_1,k_2\}=k\}} \varphi_t\{p_1/Q_1, p_2/Q_2\}[k_1, k_2].$$

3.4. Relations

Implication $\Rightarrow$ *(implies)*.

(a) In the third-valued logic:

$$\varphi_t\{p_1/Q_1 \Rightarrow p_2/Q_2\}[1] =$$

$$= \varphi_t\{p_1/Q_1, p_2/Q_2\}[0, 0] + \varphi_t\{p_1/Q_1, p_2/Q_2\}[0, 1/2]+$$
$$+\varphi_t\{p_1/Q_1, p_2/Q_2\}[0, 1] + \varphi_t\{p_1/Q_1, p_2/Q_2\}[1/2, 1/2]+$$
$$+\varphi_t\{p_1/Q_1, p_2/Q_2\}[1/2, 1] + \varphi_t\{p_1/Q_1, p_2/Q_2\}[1, 1],$$
$$\varphi_t\{p_1/Q_1 \Rightarrow p_2/Q_2\}[1/2] =$$
$$= \varphi_t\{p_1/Q_1, p_2/Q_2\}[1/2, 0] + \varphi_t\{p_1/Q_1, p_2/Q_2\}[1, 1/2],$$
$$\varphi_t\{p_1/Q_1 \Rightarrow p_2/Q_2\}[0] = \varphi_t\{p_1/Q_1, p_2/Q_2\}[1, 0].$$

(b) In the four-valued logic:

$$\varphi_t\{p_1/Q_1 \Rightarrow p_2/Q_2\}[1] =$$

$$= \varphi_t\{p_1/Q_1, p_2/Q_2\}[0, 0] + \varphi_t\{p_1/Q_1, p_2/Q_2\}[0, 1/3]+$$
$$+\varphi_t\{p_1/Q_1, p_2/Q_2\}[0, 2/3] + \varphi_t\{p_1/Q_1, p_2/Q_2\}[0, 1]+$$
$$+\varphi_t\{p_1/Q_1, p_2/Q_2\}[1/3, 1/3] + \varphi_t\{p_1/Q_1, p_2/Q_2\}[1/3, 2/3]+$$
$$+\varphi_t\{p_1/Q_1, p_2/Q_2\}[1/3, 1] + \varphi_t\{p_1/Q_1, p_2/Q_2\}[2/3, 2/3]+$$
$$+\varphi_t\{p_1/Q_1, p_2/Q_2\}[2/3, 1] + \varphi_t\{p_1/Q_1, p_2/Q_2\}[1, 1],$$
$$\varphi_t\{p_1/Q_1 \Rightarrow p_2/Q_2\}[2/3] = \varphi_t\{p_1/Q_1, p_2/Q_2\}[1/3, 0]+$$
$$+\varphi_t\{p_1/Q_1, p_2/Q_2\}[2/3, 1/3] + \varphi_t\{p_1/Q_1, p_2/Q_2\}[1, 2/3],$$
$$\varphi_t\{p_1/Q_1 \Rightarrow p_2/Q_2\}[1/3] -$$
$$= \varphi_t\{p_1/Q_1, p_2/Q_2\}[2/3, 0] + \varphi_t\{p_1/Q_1, p_2/Q_2\}[1, 1/3],$$
$$\varphi_t\{p_1/Q_1 \Rightarrow p_2/Q_2\}[0] = \varphi_t\{p_1/Q_1, p_2/Q_2\}[1, 0].$$

(c) In the m-valued logic:

$$\varphi_t\{p_1/Q_1 \Rightarrow p_2/Q_2\}[1] =$$

$$= \sum_{\{(k_1,k_2);k_1 \le k_2\}} \varphi_t\{p_1/Q_1, p_2/Q_2\}[k_1, k_2],$$

$$\varphi_t\{p_1/Q_1 \Rightarrow p_2/Q_2\}[k] =$$

$$= \sum_{\{(k_1,k_2);k_1 > k_2, 1-k_1+k_2=k\}} \varphi_t\{p_1/Q_1, p_2/Q_2\}[k_1, k_2], \quad (k \ne 1).$$

Weak equivalence $\Leftrightarrow$ *(weakly equivalent to).*

(a) In the third-valued logic:

$$\varphi_t\{p_1/Q_1 \Leftrightarrow p_2/Q_2\}[1] = \varphi_t\{p_1/Q_1, p_2/Q_2\}[0, 0]+$$

$$+\varphi_t\{p_1/Q_1, p_2/Q_2\}[1/2, 1/2] + \varphi_t\{p_1/Q_1, p_2/Q_2\}[1, 1],$$

$$\varphi_t\{p_1/Q_1 \Leftrightarrow p_2/Q_2\}[1/2] =$$

$$= \varphi_t\{p_1/Q_1, p_2/Q_2\}[0, 1/2] + \varphi_t\{p_1/Q_1, p_2/Q_2\}[1/2, 0]+$$

$$+\varphi_t\{p_1/Q_1, p_2/Q_2\}[1/2, 1] + \varphi_t\{p_1/Q_1, p_2/Q_2\}[1, 1/2],$$

$$\varphi_t\{p_1/Q_1 \Leftrightarrow p_2/Q_2\}[0] =$$

$$= \varphi_t\{p_1/Q_1, p_2/Q_2\}[0, 1] + \varphi_t\{p_1/Q_1, p_2/Q_2\}[1, 0].$$

(b) In the four-valued logic:

$$\varphi_t\{p_1/Q_1 \Leftrightarrow p_2/Q_2\}[1] =$$

$$= \varphi_t\{p_1/Q_1, p_2/Q_2\}[0, 0] + \varphi_t\{p_1/Q_1, p_2/Q_2\}[1/3, 1/3]+$$

$$+\varphi_t\{p_1/Q_1, p_2/Q_2\}[2/3, 2/3] + \varphi_t\{p_1/Q_1, p_2/Q_2\}[1, 1],$$

$$\varphi_t\{p_1/Q_1 \Leftrightarrow p_2/Q_2\}[2/3] =$$

$$= \varphi_t\{p_1/Q_1, p_2/Q_2\}[0, 1/3] + \varphi_t\{p_1/Q_1, p_2/Q_2\}[1/3, 0]+$$

$$+\varphi_t\{p_1/Q_1, p_2/Q_2\}[1/3, 2/3] + \varphi_t\{p_1/Q_1, p_2/Q_2\}[2/3, 1/3]+$$

$$+\varphi_t\{p_1/Q_1, p_2/Q_2\}[2/3, 1] + \varphi_t\{p_1/Q_1, p_2/Q_2\}[1, 2/3],$$

$$\varphi_t\{p_1/Q_1 \Leftrightarrow p_2/Q_2\}[1/3] =$$

$$= \varphi_t\{p_1/Q_1, p_2/Q_2\}[0, 2/3] + \varphi_t\{p_1/Q_1, p_2/Q_2\}[2/3, 0]+$$

$$+\varphi_t\{p_1/Q_1, p_2/Q_2\}[1/3, 1] + \varphi_t\{p_1/Q_1, p_2/Q_2\}[1, 1/3],$$

$$\varphi_t\{p_1/Q_1 \Leftrightarrow p_2/Q_2\}[0] =$$
$$= \varphi_t\{p_1/Q_1, p_2/Q_2\}[0,1] + \varphi_t\{p_1/Q_1, p_2/Q_2\}[1,0].$$

(c) In the m-valued logic:

$$\varphi_t\{p_1/Q_1 \Leftrightarrow p_2/Q_2\}[k] =$$
$$= \sum_{\{(k_1,k_2); 1-|k_1-k_2|=k\}} \varphi_t\{p_1/Q_1, p_2/Q_2\}[k_1, k_2].$$

Strong equivalence $\equiv$ *(strongly equivalent to)*.

(a) In the three-valued logic:

$$\varphi_t\{p_1/Q_1 \equiv p_2/Q_2\}[1] = \varphi_t\{p_1/Q_1, p_2/Q_2\}[0,0]+$$
$$+\varphi_t\{p_1/Q_1, p_2/Q_2\}[1/2, 1/2] + \varphi_t\{p_1/Q_1, p_2/Q_2\}[1,1],$$
$$\varphi_t\{p_1/Q_1 \equiv p_2/Q_2\}[0] =$$
$$= \varphi_t\{p_1/Q_1, p_2/Q_2\}[0, 1/2] + \varphi_t\{p_1/Q_1, p_2/Q_2\}[1/2, 0]+$$
$$+\varphi_t\{p_1/Q_1, p_2/Q_2\}[0, 1] + \varphi_t\{p_1/Q_1, p_2/Q_2\}[1,0]+$$
$$+\varphi_t\{p_1/Q_1, p_2/Q_2\}[1/2, 1] + \varphi_t\{p_1/Q_1, p_2/Q_2\}[1, 1/2].$$

(b) In the four-valued logic:

$$\varphi_t\{p_1/Q_1 \equiv p_2/Q_2\}[1] =$$
$$= \varphi_t\{p_1/Q_1, p_2/Q_2\}[0,0] + \varphi_t\{p_1/Q_1, p_2/Q_2\}[1/3, 1/3]+$$
$$+\varphi_t\{p_1/Q_1, p_2/Q_2\}[2/3, 2/3] + \varphi_t\{p_1/Q_1, p_2/Q_2\}[1,1],$$
$$\varphi_t\{p_1/Q_1 \equiv p_2/Q_2\}[0] =$$
$$= \varphi_t\{p_1/Q_1, p_2/Q_2\}[0, 1/3] + \varphi_t\{p_1/Q_1, p_2/Q_2\}[1/3, 0]+$$
$$+\varphi_t\{p_1/Q_1, p_2/Q_2\}[0, 2/3] + \varphi_t\{p_1/Q_1, p_2/Q_2\}[2/3, 0]+$$
$$+\varphi_t\{p_1/Q_1, p_2/Q_2\}[0, 1] + \varphi_t\{p_1/Q_1, p_2/Q_2\}[1, 0]+$$
$$+\varphi_t\{p_1/Q_1, p_2/Q_2\}[1/3, 2/3] + \varphi_t\{p_1/Q_1, p_2/Q_2\}[2/3, 1/3]+$$
$$+\varphi_t\{p_1/Q_1, p_2/Q_2\}[1/3, 1] + \varphi_t\{p_1/Q_1, p_2/Q_2\}[1, 1/3]+$$

$$+\varphi_t\{p_1/Q_1, p_2/Q_2\}[2/3, 1] + \varphi_t\{p_1/Q_1, p_2/Q_2\}[1, 2/3].$$

(c) In the m-valued logic:

$$\varphi_t\{p_1/Q_1 \equiv p_2/Q_2\}[1] = \sum_k \varphi_t\{p_1/Q_1, p_2/Q_2\}[k, k],$$

$$\varphi_t\{p_1/Q_1 \equiv p_2/Q_2\}[0] = \sum_{\{(k_1,k_2);k_1 \neq k_2\}} \varphi_t\{p_1/Q_1, p_2/Q_2\}[k_1, k_2].$$

Absorption.

$$\varphi_t\{p/Q, p/Q\}[k_1, k_2] = \begin{cases} \varphi_t\{p/Q\}[k], & \text{if } k_1 = k_2 = k, \\ \\ 0, & \text{if } k_1 \neq k_2. \end{cases}$$

Logical equality (at time t).

The sentences p_1, under circumstances Q_1, and p_2, under circumstances Q_2, are logically equal at time t, denoted by:

$$\varphi_t\{p_1/Q_1\} = \varphi_t\{p_2/Q_2\},$$

if and only if:

$$\varphi_t\{p_1/Q_1\}[k] = \varphi_t\{p_2/Q_2\}[k],$$

for all truth (logical) values k.

Independence (at time t).

The sentences p_1, under circumstances Q_1, and p_2, under circumstances Q_2, are independent at time t if:

$$\varphi_t\{p_1/Q_1, p_2/Q_2\}[k, \ell] = \varphi_t\{p_1/Q_1\}[k] \, \varphi_t\{p_2/Q_2\}[\ell],$$

for all truth values k, ℓ.

Independence (at times t_1 and t_2).

The sentences p_1, under circumstances Q_1, at time t_1, and p_2, under circumstances Q_2, at time t_2, are independent if:

$$\varphi_{t_1,t_2}\{p_1/Q_1, p_2/Q_2\}[k, \ell] = \varphi_{t_1}\{p_1/Q_1\}[k] \, \varphi_{t_2}\{p_2/Q_2\}[\ell],$$

for all truth values k, ℓ.

Total incompatibility (at time t).

The sentences p_1, under circumstances Q_1, and p_2, under circumstances Q_2, are totally incompatible at time t if:

$$\varphi_t\{p_1/Q_1, p_2/Q_2\}[k, k] = 0,$$

for all truth values k.

Total compatibility (at time t).

The sentences p_1, under circumstances Q_1, and p_2, under circumstances Q_2, are totally compatible at time t if:

$$\varphi_t\{p_1/Q_1, p_2/Q_2\}[k, \ell] = 0,$$

for all truth values k and ℓ such that $k \neq \ell$.

Remark: Two sentences, p_1, under circumstances Q_1, and p_2, under circumstances Q_2, totally compatible at time t, are both strongly and weakly equivalent, with certainty, at time t. Indeed, in such a case:

$$\varphi_t\{p_1/Q_1 \equiv p_2/Q_2\}[1] = \varphi_t\{p_1/Q_1 \Leftrightarrow p_2/Q_2\}[1] =$$

$$= \sum_k \varphi_t\{p_1/Q_1, p_2/Q_2\}[k, k] = \sum_{k_1, k_2} \varphi_t\{p_1/Q_1, p_2/Q_2\}[k_1, k_2] = 1.$$

Commutativity.

$$\varphi_t\{p_1/Q_1, p_2/Q_2\}[k, \ell] = \varphi_t\{p_2/Q_2, p_1/Q_1\}[\ell, k],$$

for all truth values k and ℓ. This property is assumed to be satisfied.

Marginality.

$$\varphi_t\{p_1/Q_1\}[k] = \sum_\ell \varphi_t\{p_1/Q_1, p_2/Q_2\}[k, \ell],$$

for all truth values k.

3.5. Inference

Substitution.

Let p_1 and p_2 be two atomic (simple) sentences under circumstances (contexts) Q_1 and Q_2, respectively, and let P be a molecular (complex) sentence involving (i.e., containing) the sentence p_1 under circumstances Q_1.

If we have:
$$\varphi_t\{P(p_1/Q_1)\}[k] = \alpha,$$
$$\varphi_t\{p_1/Q_1, p_2/Q_2\}[k, \ell] = 0,$$
for all k, ℓ such that $k \neq \ell$,
$$\varphi_t\{p_1/Q_1\}[\ell] = \varphi_t\{p_2/Q_2\}[\ell],$$
for all ℓ, and p_1/Q_1 and p_2/Q_2 are independent of all other atomic sentences involved in P, then:
$$\varphi_t\{P(p_2/Q_2)\}[k] = \alpha.$$

According to this rule, at any time t, we can replace p_1/Q_1 by a totally compatible and logically equal p_2/Q_2 without affecting the degree of confidence of the truth value of the complex sentence P which involved p_1/Q_1, provided that both p_1/Q_1 and p_2/Q_2 are independent of all other atomic sentences involved in P.

Modus ponens.

(a) Using the logical implication:
If we have:
$$\varphi_t\{p_1/Q_1\}[1] = \alpha,$$
$$\varphi_t\{p_1/Q_1 \Rightarrow p_2/Q_2\}[1] = \beta,$$
$$\sum_{\{(k_1,k_2);k_2<k_1<1\}} \varphi_t\{p_1/Q_1, p_2/Q_2\}[k_1, k_2] = \delta,$$
where:
$$0 \leq \alpha \leq 1, \quad 0 \leq \beta \leq 1, \quad 0 \leq \delta \leq 1,$$
$$\alpha + \beta + \delta \geq 1,$$

then:
$$\varphi_t\{p_1/Q_1, p_2/Q_2\}[1,1] = \alpha + \beta + \delta - 1.$$

Justification: Indeed, we have:

$$\alpha = \varphi_t\{p_1/Q_1\}[1] = \sum_{k_2} \varphi_t\{p_1/Q_1, p_2/Q_2\}[1, k_2],$$

$$\beta = \varphi_t\{p_1/Q_1 \Rightarrow p_2/Q_2\}[1] = \sum_{\{(k_1,k_2); k_1 \le k_2\}} \varphi_t\{p_1/Q_1, p_2/Q_2\}[k_1, k_2],$$

$$\delta = \sum_{\{(k_1,k_2); k_2 < k_1 < 1\}} \varphi_t\{p_1/Q_1, p_2/Q_2\}[k_1, k_2].$$

Summing up these three equalities and taking into account that:

$$\sum_{k_1, k_2} \varphi_t\{p_1/Q_1, p_2/Q_2\}[k_1, k_2] = 1,$$

we get the equality:

$$\alpha + \beta + \delta = \varphi_t\{p_1/Q_1, p_2/Q_2\}[1, 1] + 1,$$

because the following decomposition holds:

$$\{(k_1, k_2): k_1, k_2 \in V\} = \{(k_1, k_2): k_1, k_2 \in V, k_1 \le k_2\} \cup$$

$$\cup \{(k_1, k_2): k_1, k_2 \in V, k_2 < k_1\} = \{(k_1, k_2): k_1, k_2 \in V, k_1 \le k_2\} \cup$$

$$\cup \{(1, k_2): k_2 \in V, k_2 < 1\} \cup \{(k_1, k_2): k_1, k_2 \in V, k_2 < k_1 < 1\}. \quad \square$$

The above rule shows the relationship between the degrees of certainty of the true value of the sentence p_1/Q_1 and the true value of the logical implication $p_1/Q_1 \Rightarrow p_2/Q_2$, on one side, and the degree of certainty of the true value of the sentence p_2/Q_2, on the other side.

Remark: In particular, if $\alpha = \beta = 1$ and $\delta = 0$, then:

$$\varphi_t\{p_1/Q_1, p_2/Q_2\}[1, 1] = 1,$$

showing that if p_1/Q_1 is true, with certainty, and p_1/Q_1 logically implies p_2/Q_2, with certainty, then p_2/Q_2 is also true, with certainty.

(b) Using the probabilistic implication:
If:

$$\varphi_{t_1}\{p_1/Q_1\}[k_1]$$

and:

$$\varphi_{t_2|t_1}\{p_2/Q_2 \mid p_1/Q_1\}[k_2 \mid k_1]$$

are given, for all logical values k_1, k_2, then:

$$\varphi_{t_2}\{p_2/Q_2\}[k_2] = \sum_{k_1} \varphi_{t_2|t_1}\{p_2/Q_2 \mid p_1/Q_1\}[k_2 \mid k_1]\,\varphi_{t_1}\{p_1/Q_1\}[k_1].$$

According to this rule, knowing the degrees of certainty of the sentence p_1/Q_1 at time t_1, and of the probabilistic implication $p_2/Q_2 \mid p_1/Q_1$ at time t_2 with respect to t_1, we can calculate the degree of certainty of the sentence p_2/Q_2 at time t_2.

Remark: Let us notice that if:

$$\varphi_t\{p_2/Q_2 \mid p_1/Q_1\}[1 \mid 1] = \varphi_{t|t}\{p_2/Q_2 \mid p_1/Q_1\}[1,1] = 1,$$

and:

$$\varphi_t\{p_1/Q_1\}[1] = 1,$$

then:

$$\varphi_t\{p_2/Q_2\}[1] = 1.$$

Modus tollens:

(a) Using the logical implication:
If we have:

$$\varphi_t\{p_2/Q_2\}[0] = \alpha,$$

$$\varphi_t\{p_1/Q_1 \Rightarrow p_2/Q_2\}[1] = \beta,$$

$$\sum_{\{(k_1,k_2);0<k_2<k_1\}} \varphi_t\{p_1/Q_1, p_2/Q_2\}[k_1, k_2] = \delta,$$

where:

$$0 \le \alpha \le 1, \quad 0 \le \beta \le 1, \quad 0 \le \delta \le 1,$$

$$\alpha + \beta + \delta \ge 1,$$

then:

$$\varphi_t\{p_1/Q_1, p_2/Q_2\}[0,0] = \alpha + \beta + \delta - 1.$$

Justification: Indeed, we have:

$$\alpha = \varphi_t\{p_2/Q_2\}[0] = \sum_{k_1} \varphi_t\{p_1/Q_1, p_2/Q_2\}[k_1, 0],$$

$$\beta = \varphi_t\{p_1/Q_1 \Rightarrow p_2/Q_2\}[1] = \sum_{\{(k_1,k_2); k_1 \leq k_2\}} \varphi_t\{p_1/Q_1, p_2/Q_2\}[k_1, k_2],$$

$$\delta = \sum_{\{(k_1,k_2); 0 < k_2 < k_1\}} \varphi_t\{p_1/Q_1, p_2/Q_2\}[k_1, k_2].$$

Summing up these three equalities and taking into account that:

$$\sum_{k_1,k_2} \varphi_t\{p_1/Q_1, p_2/Q_2\}[k_1, k_2] = 1,$$

we get the equality:

$$\alpha + \beta + \delta = \varphi_t\{p_1/Q_1, p_2/Q_2\}[0,0] + 1,$$

because the following decomposition holds:

$$\{(k_1, k_2); k_1, k_2 \in V\} = \{(k_1, k_2); k_1, k_2 \in V, k_1 \leq k_2\} \cup$$

$$\cup \{(k_1, k_2); k_1, k_2 \in V, k_2 < k_1\} = \{(k_1, k_2); k_1, k_2 \in V, k_1 \leq k_2\} \cup$$

$$\cup \{(k_1, 0); k_1 \in V, k_1 > 0\} \cup \{(k_1, k_2); k_1, k_2 \in V, 0 < k_2 < k_1\}. \quad \square$$

The above rule shows the relationship between the degrees of certainty of the false value of the sentence p_2/Q_2 and the true value of the logical implication $p_1/Q_1 \Rightarrow p_2/Q_2$, on one side, and the degree of certainty of the false value of the sentence p_1/Q_1, on the other side.

Remark: In particular, if $\alpha = \beta = 1$ and $\delta = 0$, then:

$$\varphi_t\{p_1/Q_1, p_2/Q_2\}[0, 0] = 1,$$

showing that if p_2/Q_2 is false, with certainty, and p_1/Q_1 does logically imply p_2/Q_2, with certainty, then p_1/Q_1 is also false with certainty.

(b) Using the probabilistic implication:

If:

$$\varphi_{t_2}\{p_2/Q_2\}[k_2],$$

and:

$$\varphi_{t_2|t_1}\{p_2/Q_2 \mid p_1/Q_1\}[k_2 \mid k_1],$$

are given, for all logical values k_1, k_2, then:

$$\varphi_{t_1}\{p_1/Q_1\}[k_1]$$

may be determined by solving the linear system of equations:

$$\varphi_{t_2}\{p_2/Q_2\}[k_2] = \sum_{k_1} \varphi_{t_2|t_1}\{p_2/Q_2 \mid p_1/Q_1\}[k_2 \mid k_1]\,\varphi_{t_1}\{p_1/Q_1\}[k_1].$$

According to this rule, knowing the degrees of certainty of the sentence p_2/Q_2 at time t_2, and of the probabilistic implication $p_2/Q_2 \mid p_1/Q_1$, at time t_2 with respect to t_1, we can calculate the degree of certainty of the sentence p_1/Q_1 at time t_1, provided that the linear system may be solved.

As k_1 and k_2 take on the possible logical values:

$$0, \frac{1}{m-1}, \frac{2}{m-1}, \ldots, \frac{m-2}{m-1}, 1,$$

the previous system of linear equations may be written as:

$$\sum_{j=1}^{m} a_{ij}\, x_j = b_i, \quad (i = 1, \ldots, m),$$

where we have denoted by:

$$i = (m-1)k_2 + 1, \qquad j = (m-1)k_1 + 1,$$

$$a_{ij} = \varphi_{t_2|t_1}\{p_2/Q_2 \mid p_1/Q_1\}[k_2 \mid k_1],$$

$$b_i = \varphi_{t_2}\{p_2/Q_2\}[k_2], \qquad x_j = \varphi_{t_1}\{p_1/Q_1\}[k_1].$$

If the matrix $A = [a_{ij}]$ is nonsingular, i.e., its determinant $\det(A) \neq 0$, then the unique solution is:

$$x_j = (A_{1j}b_1 + \ldots A_{mj}b_m)/\det(A), \quad (j = 1, \ldots, m),$$

where A_{ij} is the cofactor of element a_{ij} in the matrix A.

Remarks: 1) Let us notice that if $t_1 = t_2 = t$, and:

$$\varphi_t\{p_2/Q_2 \mid p_1/Q_1\}[k \mid k] = \varphi_{t|t}\{p_2/Q_2 \mid p_1/Q_1\}[k \mid k] = 1,$$

for all logical values $k \in V$, and:

$$\varphi_t\{p_2/Q_2\}[0] = 1,$$

then:

$$\varphi_t\{p_1/Q_1\}[0] = 1.$$

2) We have:

$$\varphi_t\{p/Q, p/Q\}[k_1, k_2] = 0, \quad \text{if } k_1 \neq k_2,$$

but we can have:

$$\varphi_t\{p/Q_1, p/Q_2\}[k_1, k_2] > 0, \quad \text{if } k_1 \neq k_2,$$

if the circumstances Q_1 and Q_2 are not identical, or:

$$\varphi_{t_1,t_2}\{p/Q, p/Q\}[k_1, k_2] > 0, \quad \text{if } k_1 \neq k_2 \text{ and } t_1 \neq t_2.$$

3) In relative logic, it is possible to have:

$$\textit{Thesis:} \qquad \varphi_{t_1}\{p/Q_1\}[k] > 0;$$

$$\textit{Antithesis:} \qquad \varphi_{t_2}\{\sim p/Q_2\}[k] > 0;$$

$$\textit{Synthesis:} \qquad \varphi_{t_3}\{p/Q_3, \sim p/Q_4\}[k, k] > 0,$$

for all logical values k, where $t_1 \leq t_2 \leq t_3$.

Multi-valued relative logics are more diverse and flexible but they generally have fewer theorems than the classic two-valued logic. Also, it is quite often said that both the principle of contradiction and the principle of excluded third cease to be true in the multi-valued logics. Ontologically speaking, however, the principle of contradiction simply says that "Nihil idem est et non est," i.e., "Nothing *is not* and *is* at the same time," whereas the principle of excluded third states that "Quodlibet est vel non est," i.e., "Everything either *is* or *is not*," as emphasized by the scholastic logicians. Taken as such, these two principles may be universally applied to any specific domain. Thus, if we are interested in the truth of sentences, then the two principles, correctly applied, state that "No sentence is true and not true at the same time and under the same circumstances," and "At any time and under the same circumstances, every sentence is either true or is not true." In a three-valued logic with the logical values "true", "possible", and "false", the usual formulation of the principle of excluded third, i.e., "Every sentence is either true or false," is obviously not valid, but the ontological formulation of this principle continues to be correct when it is applied to each possible logical value, i.e., "At any time and under the same circumstances, every sentence is either true or is not true," "At any time and under the same circumstances, every sentence is either possible or is not possible," "At any time and under the same circumstances, every sentence is either false or is not false." The same remarks may be made about the principle of contradiction. No sentence is not true and is true, is not possible and is possible, is not false and is false, at the same time and under the same circumstances. All these considerations are drastically infirmed if time is not frozen and/or if the circumstances are allowed to change. Neither the principle of contradiction nor the principle of excluded third is universally valid even when the general, ontological form is used. The main importance of the relative logic is to highlight that the standard logical systems are extremely fragile and may be applied only when time is ignored and circumstances are considered to be invariant, in deep contradiction to the general connection, the inevitable evolution, and the ever-changing world.

If $f(p_1/Q_1, p_2/Q_2)$ is a logical function involving two arbitrary sentences p_1 and p_2 under circumstances Q_1 and Q_2, respectively, it is a tautology, or theorem, of a logical system at time t, if

$$\varphi_t\{f(p_1/Q_1, p_2/Q_2)\}[1] = 1,$$

where 1 between the square brackets is the logical value "true", often denoted by T. In a multi-valued logic with n possible logical values $\{T, F, V_1, \ldots, V_{n-2}\}$, where T and F are the values "true" and "false", respectively, $f(p_1/Q_1, p_2/Q_2)$ is a tautology at time t if its logical values are as shown in the following tableau, for all possible logical values of p_1/Q_1 and p_2/Q_2:

$f(p_1/Q_1, p_2/Q_2)$	T	F	V_1	V_2	$\ldots$	V_{n-2}
T	T	T	T	T	$\ldots$	T
F	T	T	T	T	$\ldots$	T
V_1	T	T	T	T	$\ldots$	T
V_2	T	T	T	T	$\ldots$	T
.	.	.	.	.	.	.
.	.	.	.	.	.	.
.	.	.	.	.	.	.
V_{n-2}	T	T	T	T	$\ldots$	T

In a two-valued logic, with the possible logical values $\{T, F\}$, the compound sentence $f(p_1/Q_1, p_2/Q_2)$ is a tautology (or theorem) if its logical values are as shown in the following tableau:

$f(p_1/Q_1, p_2/Q_2)$	T	F
T	T	T
F	T	T

As the last tableau is contained in the upper left corner of the first bigger tableau, we can see that any tautology in a multi-valued logic is also a tautology in the classic two-valued logic. The converse statement is not true, as shown by the two popositions given below.

Proposition 3.3: *In a two-valued relative logic, having the possible logical values $\{0, 1\}$, where 1 is "true" and 0 is "false",*

$$(\sim p/Q \Rightarrow p/Q) \Rightarrow p/Q$$

is a tautology (theorem).

Proof: We have:

$$\varphi_t\{(\sim p/Q \Rightarrow p/Q) \Rightarrow p/Q\}[0] = \varphi_t\{\sim p/Q \Rightarrow p/Q, p/Q\}[1,0] =$$

$$= \varphi_t\{\sim p/Q, p/Q, p/Q\}[0,0,0] + \varphi_t\{\sim p/Q, p/Q, p/Q\}[0,1,0] +$$

$$+\varphi_t\{\sim p/Q, p/Q, p/Q\}[1,1,0] = 0,$$

which implies:

$$\varphi_t\{(\sim p/Q \Rightarrow p/Q) \Rightarrow p/Q\}[1] = 1. \quad \square$$

Proposition 3.4: *In a three-valued relative logic, having the possible logical values $\{0, 1/2, 1\}$, where 1 is "true", $1/2$ is "possible", and 0 is "false",*

$$(\sim p/Q \Rightarrow p/Q) \Rightarrow p/Q$$

is not a tautology (theorem).

Proof: We have:

$$\varphi_t\{(\sim p/Q \Rightarrow p/Q) \Rightarrow p/Q\}[1] = \varphi_t\{\sim p/Q \Rightarrow p/Q, p/Q\}[0,0] +$$

$$+\varphi_t\{\sim p/Q \Rightarrow p/Q, p/Q\}[0,1/2] + \varphi_t\{\sim p/Q \Rightarrow p/Q, p/Q\}[0,1] +$$

$$+\varphi_t\{\sim p/Q \Rightarrow p/Q, p/Q\}[1/2,1/2] + \varphi_t\{\sim p/Q \Rightarrow p/Q, p/Q\}[1/2,1] +$$

$$+\varphi_t\{\sim p/Q \Rightarrow p/Q, p/Q\}[1,1] =$$

$$= \varphi_t\{\sim p/Q, p/Q, p/Q\}[1,0,0] + \varphi_t\{\sim p/Q, p/Q, p/Q\}[0,1,1] =$$

$$= \varphi_t\{p/Q\}[0] + \varphi_t\{p/Q\}[1],$$

whereas:

$$\varphi_t\{(\sim p/Q \Rightarrow p/Q) \Rightarrow p/Q\}[1/2] = \varphi_t\{\sim p/Q \Rightarrow p/Q, p/Q\}[1/2,0] +$$

$$+\varphi_t\{\sim p/Q \Rightarrow p/Q, p/Q\}[1,1/2] = \varphi_t\{\sim p/Q, p/Q, p/Q\}[1/2,0,0] +$$

$$+\varphi_t\{\sim p/Q, p/Q, p/Q\}[1,1/2,0] + \varphi_t\{\sim p/Q, p/Q, p/Q\}[0,0,1/2] +$$

$$+\varphi_t\{\sim p/Q, p/Q, p/Q\}[0,1/2,1/2] + \varphi_t\{\sim p/Q, p/Q, p/Q\}[0,1,1/2] +$$

$$+\varphi_t\{\sim p/Q, p/Q, p/Q\}[1/2,1/2,1/2] + \varphi_t\{\sim p/Q, p/Q, p/Q\}[1/2,1,1/2] +$$

$$+\varphi_t\{\sim p/Q, p/Q, p/Q\}[1,1,1/2] = \varphi_t\{\sim p/Q, p/Q, p/Q\}[1/2,1/2,1/2] =$$

$$= \varphi_t\{p/Q, p/Q, p/Q\}[1/2, 1/2, 1/2] = \varphi_t\{p/Q\}[1/2],$$

and, similarly:

$$\varphi_t\{(\sim p/Q \Rightarrow p/Q) \Rightarrow p/Q\}[0] = \varphi_t\{\sim p/Q \Rightarrow p/Q, p/Q\}[1, 0] =$$

$$= \varphi_t\{\sim p/Q, p/Q, p/Q\}[0, 0, 0] + \varphi_t\{\sim p/Q, p/Q, p/Q\}[0, 1/2, 0] +$$

$$+ \varphi_t\{\sim p/Q, p/Q, p/Q\}[0, 1, 0] + \varphi_t\{\sim p/Q, p/Q, p/Q\}[1/2, 1/2, 0] +$$

$$+ \varphi_t\{\sim p/Q, p/Q, p/Q\}[1/2, 1, 0] + \varphi_t\{\sim p/Q, p/Q, p/Q\}[1, 1, 0] = 0. \quad \square$$

We can introduce two binary operators corresponding to *possibility*, denoted by M, and *doubt*, denoted by D.

Possibility M.

(a) In the three-valued logic:

$$\varphi_t\{Mp/Q\}[1] = \varphi_t\{p/Q\}[1/2] + \varphi_t\{p/Q\}[1],$$

$$\varphi_t\{Mp/Q\}[0] = \varphi_t\{p/Q\}[0].$$

(b) In the four-valued logic:

$$\varphi_t\{Mp/Q\}[1] = \varphi_t\{p/Q\}[1/3] +$$

$$+ \varphi_t\{p/Q\}[2/3] + \varphi_t\{p/Q\}[1],$$

$$\varphi_t\{Mp/Q\}[0] = \varphi_t\{p/Q\}[0].$$

(c) In the m-valued logic:

$$\varphi_t\{Mp/Q\}[1] = \sum_{\{k; k \neq 0\}} \varphi_t\{p/Q\}[k],$$

$$\varphi_t\{Mp/Q\}[0] = \varphi_t\{p/Q\}[0].$$

As we look at Mp/Q to be the possibility of sentence p under the circumstances (context) Q, $\sim M \sim p/Q$ may be interpreted as being the *necessity* of p under circumstances Q.

Remark: The binary operator $\sim M \sim$ satisfies:

$$\varphi_t\{\sim M \sim p/Q\}[1] = \varphi_t\{M \sim p/Q\}[0] =$$

$$= \varphi_t\{\sim p/Q\}[0] = \varphi_t\{p/Q\}[1],$$

$$\varphi_t\{\sim M \sim p/Q\}[0] = \varphi_t\{M \sim p/Q\}[1] =$$

$$= \sum_{\{k;k\neq 0\}} \varphi_t\{\sim p/Q\}[k] = \sum_{\{k;k\neq 0\}} \varphi_t\{p/Q\}[1-k].$$

Doubt D.

(a) In the three-valued logic:

$$\varphi_t\{Dp/Q\}[1] = \varphi_t\{p/Q\}[1/2],$$

$$\varphi_t\{Dp/Q\}[0] = \varphi_t\{p/Q\}[0] + \varphi_t\{p/Q\}[1].$$

(b) In the four-valued logic:

$$\varphi_t\{Dp/Q\}[1] = \varphi_t\{p/Q\}[1/3] + \varphi_t\{p/Q\}[2/3],$$

$$\varphi_t\{Dp/Q\}[0] = \varphi_t\{p/Q\}[0] + \varphi_t\{p/Q\}[1].$$

(c) In the m-valued logic:

$$\varphi_t\{Dp/Q\}[1] = \sum_{\{k;k\notin\{0,1\}\}} \varphi_t\{p/Q\}[k],$$

$$\varphi_t\{Dp/Q\}[0] = \varphi_t\{p/Q\}[0] + \varphi_t\{p/Q\}[1].$$

Remark: In the three-valued relative logic:

$$\varphi_t\{(p/Q) \equiv (\sim p/Q)\}[1] = \varphi_t\{p/Q, \sim p/Q\}[1/2, 1/2] =$$

$$= \varphi_t\{p/Q, p/Q\}[1/2, 1/2] = \varphi_t\{p/Q\}[1/2] = \varphi_t\{Dp/Q\}[1].$$

In a multi-valued logic we can check whether a certain logical expression is or is not a theorem (tautology), as we did in a two-valued relative logic in Chapter 2. The next proposition is typical from this point of view and we can try a similar approach in dealing with any logical formula:

Proposition 3.5: *In an m-valued relative logic, the expression:*

$$p_1/Q_1 \Rightarrow (p_2/Q_2 \Rightarrow p_1/Q_1)$$

is a theorem (tautology).

Proof. We have:

$$\varphi_t\{p_1/Q_1 \Rightarrow (p_2/Q_2 \Rightarrow p_1/Q_1)\}[1] =$$

$$= \sum_{\{(k_1,k);k_1\leq k\}} \varphi_t\{p_1/Q_1, p_2/Q_2 \Rightarrow p_1/Q_1\}[k_1,k] =$$

$$= \sum_{\{k_1;k_1\leq 1\}} \varphi_t\{p_1/Q_1, p_2/Q_2 \Rightarrow p_1/Q_1\}[k_1,1]+$$

$$+ \sum_{\{(k_1,k);k_1\leq k<1\}} \varphi_t\{p_1/Q_1, p_2/Q_2 \Rightarrow p_1/Q_1\}[k_1,k] =$$

$$= \varphi_t\{p_2/Q_2 \Rightarrow p_1/Q_1\}[1]+$$

$$+ \sum_{\{(k_1,k);k_1\leq k<1\}} \sum_{\{(k_2,k_3);k_2>k_3,1-k_2+k_3=k\}} \varphi_t\{p_1/Q_1, p_2/Q_2, p_1/Q_1\}[k_1,k_2,k_3] =$$

$$= \sum_{\{(k_2,k_1);k_2\leq k_1\}} \varphi_t\{p_2/Q_2, p_1/Q_1\}[k_2,k_1]+$$

$$+ \sum_{\{(k_2,k_1);k_1\leq k<1,k_2>k_1\}} \varphi_t\{p_2/Q_2, p_1/Q_1\}\}[k_2,k_1] =$$

$$= \sum_{\{(k_2,k_1);k_2\leq k_1\}} \varphi_t\{p_2/Q_2, p_1/Q_1\}[k_2,k_1]+$$

$$+ \sum_{\{(k_2,k_1);k_1<k_2\}} \varphi_t\{p_2/Q_2, p_1/Q_1\}[k_2,k_1] - 1,$$

because we must have $k_3 = k_1$, whereas:

$$k_1 \leq k < 1 \quad \text{and} \quad k = 1 - k_2 + k_3 = 1 - k_2 + k_1$$

$$\text{imply} \quad k_2 \leq 1 \quad \text{and} \quad k_2 > k_1. \quad \square$$

Proposition 3.6: *In an m-valued relative logic, the implication:*

$$p/Q \Rightarrow Mp/Q$$

is a theorem (tautology).

Proof. We have:

$$\varphi_t\{p/Q \Rightarrow Mp/Q\}[1] = \sum_{\{(k_1,k_2);k_1\leq k_2\}} \varphi_t\{p/Q, Mp/Q\}[k_1, k_2] =$$

$$= \varphi_t\{p/Q, Mp/Q\}[0, 0] + \sum_{\{k_1;k_1\leq 1\}} \varphi_t\{p/Q, Mp/Q\}[k_1, 1] =$$

$$= \varphi_t\{p/Q, p/Q\}[0, 0] + \varphi_t\{Mp/Q\}[1] =$$

$$= \varphi_t\{p/Q\}[0] + \varphi_t\{Mp/Q\}[1] =$$

$$= \varphi_t\{Mp/Q\}[0] + \varphi_t\{Mp/Q\}[1] = 1. \quad \square$$

Proposition 3.7: *In an m-valued relative logic, the implication:*

$$Dp/Q \Rightarrow Mp/Q$$

is a theorem (tautology), showing that doubtful implies possible.

Proof. We have:

$$\varphi_t\{Dp/Q \Rightarrow Mp/Q\}[1] = \sum_{\{(k_1,k_2);k_1\leq k_2\}} \varphi_t\{Dp/Q, Mp/Q\}[k_1, k_2] =$$

$$= \varphi_t\{Dp/Q, Mp/Q\}[0, 0] + \varphi_t\{Dp/Q, Mp/Q\}[0, 1]+$$

$$+\varphi_t\{Dp/Q, Mp/Q\}[1, 1] = \varphi_t\{p/Q, p/Q\}[0, 0]+$$

$$+\varphi_t\{p/Q, p/Q\}[1, 0] + \sum_{\{k;k\neq 0\}} \varphi_t\{p/Q, p/Q\}[0, k]+$$

$$+ \sum_{\{k;k\neq 0\}} \varphi_t\{p/Q, p/Q\}[1, k]+$$

$$+ \sum_{\{(k_1,k_2);k_1\neq 0,k_1\neq 1,k_2\neq 0\}} \varphi_t\{p/Q, p/Q\}[k_1, k_2] =$$

$$= \varphi_t\{p/Q\}[0] + \varphi_t\{p/Q\}[1] + \sum_{\{k_1; k_1 \neq 0, k_1 \neq 1\}} \varphi_t\{p/Q\}[k_1] =$$

$$= \sum_k \varphi_t\{p/Q\}[k] = 1. \quad \square$$

Proposition 3.8: *In an m-valued relative logic, the equivalence:*

$$p/Q \equiv\sim (\sim p/Q)$$

is a theorem (tautology).

Proof. We have:

$$\varphi_t\{p/Q \equiv\sim (\sim p/Q)\}[1] = \sum_k \varphi_t\{p/Q, \sim (\sim p/Q)\}[k, k] =$$

$$= \sum_k \varphi_t\{p/Q, \sim p/Q\}[k, 1-k] = \sum_k \varphi_t\{p/Q, p/Q\}[k, k] =$$

$$= \sum_k \varphi_t\{p/Q\}[k] = 1. \quad \square$$

A stronger result is given in the next proposition:

Proposition 3.9. *In an m-valued relative logic, the logical equality:*

$$\varphi_t\{\sim (\sim p/Q)\} = \varphi_t\{p/Q\}$$

holds (the law of double negation).

Proof. For every logical value k we have:

$$\varphi_t\{\sim (\sim p/Q)\}[k] = \varphi_t\{\sim p/Q\}[1-k] = \varphi_t\{p/Q\}[k]. \quad \square$$

A theorem is a logical expression that is true with certainty whatever the logical values of the sentences making up the respective expression are. It is equally important to see what happens with the probability of different logical values of an expression that is not necessarily a theorem.

Proposition 3.10: *In an m-valued relative logic, when m is an odd number,*

$$\varphi_t\{p/Q \vee (\sim p/Q)\}[k]$$

is equal to:

$$0, \quad if \quad k < 0.5,$$

$$\varphi_t\{p/Q\}[0.5], \quad if \quad k = 0.5,$$

$$\varphi_t\{p/Q\}[k] + \varphi_t\{p/Q\}[1-k], \quad if \quad k > 0.5.$$

In an m-valued logic, when m is an even number,

$$\varphi_t\{p/Q \vee (\sim p/Q)\}[k]$$

is equal to:

$$0, \quad if \quad k < 0.5,$$

$$\varphi_t\{p/Q\}[k] + \varphi_t\{p/Q\}[1-k], \quad if \quad k > 0.5.$$

Proof. Assume that m is an even number. If $k > 0.5$, we have:

$$\varphi_t\{p/Q \vee (\sim p/Q)\}[k] =$$

$$= \sum_{\{(k_1,k_2);\max\{k_1,k_2\}=k\}} \varphi_t\{p/Q, \sim p/Q\}[k_1, k_2] =$$

$$= \sum_{\{k_1;\max\{k_1,1-k_1\}=k\}} \varphi_t\{p/Q, \sim p/Q\}[k_1, 1-k_1] =$$

$$= \varphi_t\{p/Q, \sim p/Q\}[k, 1-k] + \varphi_t\{p/Q, \sim p/Q\}[1-k, k] =$$

$$= \varphi_t\{p/Q, p/Q\}[k, k] + \varphi_t\{p/Q, p/Q\}[1-k, 1-k] =$$

$$= \varphi_t\{p/Q\}[k] + \varphi_t\{p/Q\}[1-k].$$

On the other hand, if $k < 0.5$, as $1 - k > 0.5$, we get:

$$\varphi_t\{p/Q \vee (\sim p/Q)\}[k] = 0.$$

If m is an odd number, we also have:

$$\varphi_t\{p/Q \vee (\sim p/Q)\}[0.5] =$$

$$= \sum_{\{(k_1,k_2);\max\{k_1,k_2\}=0.5\}} \varphi_t\{p/Q, \sim p/Q\}[k_1, k_2] =$$

$$= \sum_{\{k_1;\max\{k_1,1-k_1\}=0.5\}} \varphi_t\{p/Q, \sim p/Q\}[k_1, 1-k_1] =$$

$$= \varphi_t\{p/Q, \sim p/Q\}[0.5, 0.5] =$$
$$= \varphi_t\{p/Q, p/Q\}[0.5, 0.5] = \varphi_t\{p/Q\}[0.5]. \quad \square$$

Remark: If we have only two logical values, namely, $V = \{0, 1\}$, we have $m = 2$ and the above formulas become:

$$\varphi_t\{p/Q \vee (\sim p/Q)\}[0] = 0,$$

$$\varphi_t\{p/Q \vee (\sim p/Q)\}[1] = \varphi_t\{p/Q\}[1] + \varphi_t\{p/Q\}[0] = 1,$$

which shows that $p/Q \vee (\sim p/Q)$ is a theorem in the two-valued logic.

Proposition 3.11: *In an m-valued relative logic, when m is an odd number,*

$$\varphi_t\{p/Q \cdot (\sim p/Q)\}[k]$$

is equal to:

$$0, \quad if \ k > 0.5,$$

$$\varphi_t\{p/Q\}[0.5], \quad if \ k = 0.5,$$

$$\varphi_t\{p/Q\}[k] + \varphi_t\{p/Q\}[1 - k], \quad if \ k < 0.5.$$

In an m-valued logic, when m is an even number,

$$\varphi_t\{p/Q \cdot (\sim p/Q)\}[k]$$

is equal to:

$$0, \quad if \ k > 0.5,$$

$$\varphi_t\{p/Q\}[k] + \varphi_t\{p/Q\}[1 - k], \quad if \ k < 0.5.$$

Proof: We have:

$$\varphi_t\{p/Q \cdot (\sim p/Q)\}[k] = \sum_{\{(k_1,k_2); \min\{k_1,k_2\}=k\}} \varphi_t\{p/Q, \sim p/Q\}[k_1, k_2].$$

If m is an even number, as $k < 0.5$ is equivalent to $1 - k > 0.5$ and $k > 0.5$ is equivalent to $1 - k < 0.5$, from the above equality, we have:

$$\varphi_t\{p/Q \cdot (\sim p/Q)\}[k] = 0,$$

if $k > 0.5$, and:

$$\varphi_t\{p/Q \cdot (\sim p/Q)\}[k] =$$

$$= \varphi_t\{p/Q, \sim p/Q\}[k, 1-k] + \varphi_t\{p/Q, \sim p/Q\}[1-k, k] =$$

$$= \varphi_t\{p/Q, p/Q\}[k, k] + \varphi_t\{p/Q, p/Q\}[1-k, 1-k] =$$

$$= \varphi_t\{p/Q\}[k] + \varphi_t\{p/Q\}[1-k],$$

if $k < 0.5$. If m is an odd number, we also have:

$$\varphi_t\{p/Q \cdot (\sim p/Q)\}[0.5] = \varphi_t\{p/Q, \sim p/Q\}[0.5, 0.5] =$$

$$= \varphi_t\{p/Q, p/Q\}[0.5, 0.5] = \varphi_t\{p/Q\}[0.5]. \quad \square$$

Remark: If we have only two logical values, namely, $V = \{0, 1\}$, we have $m = 2$ and the above formulas become:

$$\varphi_t\{p/Q \cdot (\sim p/Q)\}[1] = 0,$$

$$\varphi_t\{p/Q \cdot (\sim p/Q)\}[0] = \varphi_t\{p/Q\}[0] + \varphi_t\{p/Q\}[1] = 1,$$

respectively, showing that $p/Q \cdot (\sim p/Q)$ is a theorem in the two-valued relative logic.

Complementary remarks:

1) In 1951, G.H. von Wright constructed the *deontic logic*, i.e., the logic of duty or obligation, based on classic logic. Lothar Philipps (1964, 1966) built the deontic logic based on Heyting's intuitionist logic. This is a practical logic containing a large spectrum of logics, as the logic of values, the logic of norms, the logic of imperatives, the logic of choice, the preferential logic, the logic of better, the deviant logic, etc. The common strategy consists in defining the possible logic values and the rules for negation, disjunction, product, and implication. There are many ways of defining them and, consequently, there are many such logical systems. Very often, the independence, noncontradiction, and completeness of the axioms proposed for such practical logics are open to debate and controversy. The following table shows the possible values or levels of truth, knowledge, duty, and existence used more frequently.

Truth	*Knowledge*	*Duty*	*Existence*
necessary	verified	obligatory	universal
possible		permitted	existing
contingent	undecided	indifferent	
impossible	falsified	forbidden	empty

2) G. Birkhoff and J. von Neumann (1936) showed for the first time that the experimental sentences of quantum mechanics do not follow the rules of the two-valued logic. Paulette Février (1937) used the three-valued logic in dealing with Heisenberg's indeterminacy relations from quantum mechanics. As it is well-known, W. Heisenberg (1927) stated that the exact values of the position q and impuls p of a microparticle cannot be known simultaneously. Both quantities are experimentally determined with some errors, $\triangle q$ and $\triangle p$, respectively. Heisenberg showed that they are not independent in the subatomic world and satisfy the indeterminacy inequality:

$$\triangle p \triangle q \geq h,$$

where h is a small constant. Consequently, two experimental sentences about the position and impuls of an elementary particle at a given moment, namely, "q has the value $q_0 \pm \triangle q$" and "p has the value $p_0 \pm \triangle p$" cannot be true or false simultaneously if $\triangle p \triangle q < h$. P. Février grouped the pair of sentences in two classes: the class of pairs of sentences for which the sum and product of the classic logic may be used and the class of pairs of sentences for which the logic product is always absolutely false, a third logic value, added to the values true and false of classic logic.

In dealing with the quantum mechanics, H. Reichenbach (1944) also used a logic with three values: "true", denoted by w (from the German word "Wahrheit"), "false", denoted by f (from the German word "Falschheit"), and "undeterminable", denoted by u (from the German "Unbestimmtheit"). He introduced three negations: the "cyclic" negation, denoted by $-$, the "diametrical" negation, denoted by $\sim$, and the "complete" negation, denoted by a bar, defined by the following tableaus:

p	$-p$		p	$\sim p$		p	$\bar{p}$
w	u		w	f		w	u
u	f		u	u		u	w
f	w		f	w		f	w

The "alternative" disjunction and implication are defined by:

p	q	$p \vee q$	$p \to q$
w	w	w	w
w	u	w	f
w	f	w	f
u	w	w	w
u	u	u	w
u	f	u	w
f	w	w	w
f	u	u	w
f	f	f	w

Quantum mechanics has introduced the essential notion of "complementarity", involving two complementary descriptions of a microscopic object, like the corpuscular description and the undulatory description of an electron, for instance, that cannot be incorporated into the same unique description. If one description makes sense, i.e., it is true or false, the other one does not make sense. We simply cannot refer to both descriptions at the same time and under the same conditions. Any logic claiming to deal with quantum mechanics has to incorporate this kind of complementarity. According to Reichenbach, two propositions p and q are "complementary" if the following implication is true:

$$(p \vee -p) \to - - q.$$

As $- - q$ is equivalent to saying that q is undeterminable, the above implication says that if p is true or false then q is undeterminable. There is a symmetry between two complementary propositions p and q, expressed by the alternative implication:

$$\overline{- - p} \to - - q.$$

3) D.A. Bocivar (1938, 1943) constructed a three-valued logic for explaining the origin of paradoxes in mathematical logic. He made a distinction between statement and sentence. A statement has a more general character and may be even senseless, whereas a sentence is a particular statement which has a sense. Consequently, statements may have three values: true, false, and nonsense. By contrast, sentences may have only two values: true and false. Bocivar showed that in his system the paradoxal statements had

the nonsense value. M. Shaw-Kwei (1954), on the other hand, showed that there are paradoxes in the multi-valued systems as well.

4) L.E. Brower rejected the general valability of the principle of excluded third. He believed that the paradoxes had occurred in logic due to an indiscriminate use of this principle. In the intuitionist logic, however, this principle is not declared false but only unproved. Intuitionist logic may be viewed as a multi-valued logic. A. Heyting introduced three values for sentences: 0 for "true", 1 for "false", and 2 for "cannot be false but its truth cannot be proved." The rules for negation $\perp$, implication $\supset$ ($p_1 \supset p_2$ says that "from p_1 it results p_2"), conjunction $\wedge$, and disjunction $\vee$ are given in the following tableaus:

p	0	1	2
$\perp p$	1	0	1

$p_1 \supset p_2$	0	1	2
0	0	1	2
1	0	0	0
2	0	1	0

$p_1 \wedge p_2$	0	1	2
0	0	1	2
1	1	1	1
2	2	1	2

$p_1 \vee p_2$	0	1	2
0	0	0	0
1	0	1	2
2	0	2	2

The corresponding rules in relative logic would be:

Negation.

$$\varphi_t\{\perp p/Q\}[0] = \varphi_t\{p/Q\}[1],$$

$$\varphi_t\{\perp p/Q\}[1] = \varphi_t\{p/Q\}[0] + \varphi_t\{p/Q\}[2].$$

Implication.

$$\varphi_t\{p_1/Q_1 \supset p_2/Q_2\}[0] =$$

$$= \varphi_t\{p_1/Q_1, p_2/Q_2\}[0,0] + \varphi_t\{p_1/Q_1, p_2/Q_2\}[1,0]+$$

$$+\varphi_t\{p_1/Q_1, p_2/Q_2\}[1,1] + \varphi_t\{p_1/Q_1, p_2/Q_2\}[1,2]+$$

$$+\varphi_t\{p_1/Q_1, p_2/Q_2\}[2,0] + \varphi_t\{p_1/Q_1, p_2/Q_2\}[2,2],$$

$$\varphi_t\{p_1/Q_1 \supset p_2/Q_2\}[1] =$$

$$= \varphi_t\{p_1/Q_1, p_2/Q_2\}[0,1] + \varphi_t\{p_1/Q_1, p_2/Q_2\}[2,1],$$

$$\varphi_t\{p_1/Q_1 \supset p_2/Q_2\}[2] = \varphi_t\{p_1/Q_1, p_2/Q_2\}[0,2].$$

Conjunction.

$$\varphi_t\{p_1/Q_1 \wedge p_2/Q_2\}[0] = \varphi_t\{p_1/Q_1, p_2/Q_2\}[0,0],$$

$$\varphi_t\{p_1/Q_1 \wedge p_2/Q_2\}[1] = \varphi_t\{p_1/Q_1, p_2/Q_2\}[0,1]+$$

$$+\varphi_t\{p_1/Q_1, p_2/Q_2\}[1,0] + \varphi_t\{p_1/Q_1, p_2/Q_2\}[1,1]+$$

$$+\varphi_t\{p_1/Q_1, p_2/Q_2\}[1,2] + \varphi_t\{p_1/Q_1, p_2/Q_2\}[2,1],$$

$$\varphi_t\{p_1/Q_1 \wedge p_2/Q_2\}[2] = \varphi_t\{p_1/Q_1, p_2/Q_2\}[0,2]+$$

$$+\varphi_t\{p_1/Q_1, p_2/Q_2\}[2,0] + \varphi_t\{p_1/Q_1, p_2/Q_2\}[2,2].$$

Disjunction.

$$\varphi_t\{p_1/Q_1 \vee p_2/Q_2\}[0] = \varphi_t\{p_1/Q_1, p_2/Q_2\}[0,0]+$$

$$+\varphi_t\{p_1/Q_1, p_2/Q_2\}[0,1] + \varphi_t\{p_1/Q_1, p_2/Q_2\}[0,2]+$$

$$+\varphi_t\{p_1/Q_1, p_2/Q_2\}[1,0] + \varphi_t\{p_1/Q_1, p_2/Q_2\}[2,0],$$

$$\varphi_t\{p_1/Q_1 \vee p_2/Q_2\}[1] = \varphi_t\{p_1/Q_1, p_2/Q_2\}[1,1],$$

$$\varphi_t\{p_1/Q_1 \vee p_2/Q_2\}[2] = \varphi_t\{p_1/Q_1, p_2/Q_2\}[1,2]+$$

$$+\varphi_t\{p_1/Q_1, p_2/Q_2\}[2,1] + \varphi_t\{p_1/Q_1, p_2/Q_2\}[2,2].$$

Question: Is $p/Q \vee \perp p/Q$ a tautology (theorem) of this intuitionist logical system?

Answer: No, because:

$$\varphi_t\{p/Q \vee \perp p/Q\}[0] = \varphi_t\{p/Q, \perp p/Q\}[0,1]+$$

$$+\varphi_t\{p/Q, \perp p/Q\}[1,0] = \varphi_t\{p/Q\}[0] + \varphi_t\{p/Q\}[1],$$
$$\varphi_t\{p/Q \vee \perp p/Q\}[1] = 0,$$
$$\varphi_t\{p/Q \vee \perp p/Q\}[2] = \varphi_t\{p/Q, \perp p/Q\}[2,1] =$$
$$= \varphi_t\{p/Q\}[2]. \quad \square$$

Question: Is $\perp\perp p/Q \supset p/Q$ a tautology (theorem) of this intuitionist logical system?

Answer: No, because, taking the values of p/Q into account, we have:

$$\varphi_t\{\perp\perp p/Q \supset p/Q\}[0] = \varphi_t\{\perp\perp p/Q, p/Q\}[0,0]+$$

$$+\varphi_t\{\perp\perp p/Q, p/Q\}[1,1] = \varphi_t\{p/Q\}[0] + \varphi_t\{p/Q\}[1].$$

In the combinations that give the value 0 for the implication $\supset$ only $[0,0]$ and $[1,1]$ are valid. For the value 2 of p/Q, the pairs $[1,2]$ and $[2,2]$ are not valid. We cannot assume that $\perp\perp p/Q$ is p/Q, but if we take a logical value of p/Q, we know the logical value of $\perp p/Q$ and, afterwards, the logical value of $\perp\perp p/Q$. Similarly, we get:

$$\varphi_t\{\perp\perp p/Q \supset p/Q\}[1] = 0,$$

$$\varphi_t\{\perp\perp p/Q \supset p/Q\}[2] = \varphi_t\{\perp\perp p/Q, p/Q\}[0,2] =$$
$$= \varphi_t\{p/Q\}[2]. \quad \square$$

Question: Is $p/Q \supset (p/Q \wedge p/Q)$ a tautology (theorem) of the intuitionist logical system?

Answer: Yes, because:

$$\varphi_t\{p/Q \supset (p/Q \wedge p/Q)\}[0] = \varphi_t\{p/Q, p/Q \wedge p/Q\}[0,0]+$$

$$+\varphi_t\{p/Q, p/Q \wedge p/Q\}[1,1] + \varphi_t\{p/Q, p/Q \wedge p/Q\}[2,2] = 1,$$
$$\varphi_t\{p/Q \supset (p/Q \wedge p/Q)\}[1] = 0,$$

$$\varphi_t\{p/Q \supset (p/Q \wedge p/Q)\}[2] = 0. \quad \square$$

Question: Is $p/Q \supset \bot\bot p/Q$ a tautology (theorem) of the intuitionist logical system?

Answer: Yes, because:

$$\varphi_t\{p/Q \supset \bot\bot p/Q\}[0] = \varphi_t\{p/Q, \bot\bot p/Q\}[0,0]+$$

$$+\varphi_t\{p/Q, \bot\bot p/Q\}[1,1] + \varphi_t\{p/Q, \bot\bot p/Q\}[2,0] =$$

$$= \varphi_t\{p/Q\}[0] + \varphi_t\{p/Q\}[1] + \varphi_t\{p/Q\}[2] = 1,$$

$$\varphi_t\{p/Q \supset \bot\bot p/Q\}[1] = 0,$$

$$\varphi_t\{p/Q \supset \bot\bot p/Q\}[2] = 0. \quad \square$$

Chapter 4. STOCHASTIC LOGIC

4.1. Stochastic sets

Cantor's set theory, often called the theory of crisp sets, is intimately related to the classic logic and there is a well-known correspondence between their operations. Thus the complement, intersection, union, and inclusion from set theory correspond to the negation "non", conjunction "and", disjunction "or", and implication from classic logic. Set theory allows us to make the concepts of classic logic more intuitive. Venn diagrams, for instance, are set theory models that are frequently used to facilitate and visualize the proofs of the main results of classic logic. By analogy, stochastic sets are introduced in order to bring stochastic logic closer to the general concepts and operations from set theory.

Stochastic logic is a probabilistic logic that essentially takes interdependence and global connection into account. Sentences, properties, or predicates will refer here to the elements of a set of entities and the emphasis will be put on making these sentences true, or satisfying the corresponding properties or predicates, an alternative denoted by 1, or not, a complementary alternative denoted by 0. Therefore, this chapter is dealing with a two-valued relative logic where sentences are referred to the elements of crisp sets.

Stochastic sets play for stochastic logic the same role as that played by Cantor's sets for classic logic. In order to simplify the presentation, we are going to deal mainly with two-valued stochastic sets involving a finite number of entities. This means that the properties or predicates considered here take on only two possible values, namely 1, meaning yes (or satisfied), and 0, meaning no (or not satisfied). Let us emphasize, however, that although a property p can have only two possible values, the degree of truth of the statement "the entity x satisfies (has) property p" may be an arbitrary number from the numerical interval $[0, 1]$. Generalizations of stochastic sets to cases involving multi-valued properties or predicates and an infinite set of entities will be discussed at the end of the chapter.

Stochastic logic is a nonstandard logic, a special case of relative logic. It is not a rival of classic logic but rather a generalization of both classic and fuzzy logics. It is both probabilistic, context depending, and global. Stochastic logic is a supplementary logic in the sense that it is compatible both to classic logic, which is nonprobabilistic, or strictly deterministic, and to fuzzy logic, which may be viewed globally as being a probabilistic logic with independent components.

The probabilistic feature of the proposed logic is motivated by the necessity of coping with uncertainty in dealing with real life problems. Such uncertainty could have objective or subjective causes. Sometimes, the qualifications for being something are imprecise. Sometimes, the qualifications for being something else are precise, but there is real difficulty in determining whether or not certain subjects satisfy them. Stochastic logic, however, should not be identified to the standard probabilistic logic where the logical concept of probability replaces the two truth values (true, false) from the classic logic. Its aim is mainly to take interdependence and global connection into account. Interdependence is universal and essentially influences the values of truth. When there is interdependence among entities, the degree of truth with which one of these entities satisfies a certain property depends in general on the degrees of truth with which the other entities satisfy that property or some other properties. Sometimes this interdependence is relatively weak and may be ignored, but when it is strong, such a simplification would distort the reality.

The aim of this chapter is to present basic properties of stochastic logic and the relationship between its concepts and the standard ones. The different probability distributions involved may be objective, i.e., based either on stable relative frequencies determined by repeating probabilistic experiments, or subjective, based on the principle of maximum uncertainty, in which case they are rather called credibility distributions. As an example, the stochastic sets of murder suspects having motives and opportunity from an Agatha Christie's detective story are presented.

4.2. The formalism

Basic joint stochastic set.

Let $X = \{x_1, \ldots, x_m\}, Y = \{y_1, \ldots, y_n\}, \ldots$ be finite, not necessarily different, crisp sets of distinct *entities*. These sets are called *universes*. Let $A, B, \ldots$ be some sentences representing *properties* or *predicates*, not necessarily different. The *basic joint stochastic set* $A/X \otimes B/Y \otimes \cdots$ is completely characterized by a joint membership probability (credibility) distribution

$\chi_{A/X,B/Y,...}$ on the binary crisp product set $\{0,1\}^m \times \{0,1\}^n \times \cdots$, showing the probability (credibility) that the entities of the given universes jointly satisfy or not the respective properties. If $m = 3$ and $Y = X$, for instance, then $\chi_{A/X,B/X}(1,0,0;1,0,1)$ is the probability (credibility) that entity x_1 satisfies both A and B, *and* x_2 does not satisfy A and B, *and* x_3 satisfies B but not A. Using the notations from chapters 2 and 3, we are dealing here with a relative logic where time is ignored or is assigned to properties:

$$\chi_{A/X,B/Y,...}(i_1, \ldots, i_m, j_1, \ldots, j_n, \ldots) =$$

$$= \varphi\{A/\{x_1\}, \ldots, A/\{x_m\}, B/\{y_1\}, \ldots, B/\{y_n\}, \ldots\}[i_1, \ldots, i_m, j_1, \ldots, j_n, \ldots],$$

where each i_k and j_ℓ may be either 1 or 0, representing "yes" ("true") or "no" ("false"), respectively.

Restricted stochastic set.

The *restriction A/X* of the basic joint stochastic set $A/X \otimes B/Y \otimes \cdots$ to property A and universe X is defined by the marginal membership probability distribution $\chi_{A/X}$ obtained by saturating all the binary arguments of the joint membership probability distribution $\chi_{A/X,B/Y,...}$ except the first m ones, i.e.,

$$\chi_{A/X}(i_1, \ldots, i_m) = \sum_{j_1,\ldots,j_n,\ldots \in \{0,1\}} \chi_{A/X,B/Y,...}(i_1, \ldots, i_m, j_1, \ldots, j_n; \ldots).$$

The restriction of the stochastic set A/X to the crisp subset X^* of the universe X has as membership probability (credibility) distribution the marginal probability (credibility) distribution of $\chi_{A/X}$ relative to the elements of X^*. Thus, for instance,

$$\chi_{A/\{x_i\}}(k_i) = \sum_{k_1,\ldots,k_{i-1},k_{i+1},\ldots,k_m \in \{0,1\}} \chi_{A/\{x_1,\ldots,x_m\}}(k_1, \ldots, k_m)$$

represents the probability (credibility) that x_i has (if $k_i = 1$) or has not (if $k_i = 0$) property A regardless of what happens with the other entities of X. It defines the restriction of the membership probability (credibility) distribution $\chi_{A/X}$ to the subset $X^* = \{x_i\} \subset X$. The *complement* with respect to X of the stochastic set A/X^*, where $X^* \subset X$, is the stochastic set $A/(X-X^*)$. Thus, for instance, the complement of $A/\{x_i\}$ with respect to X

is $A/\{x_1, \ldots, x_{i-1}, x_{i+1}, \ldots, x_m\}$ whose membership probability (credibility) distribution is:

$$\chi_{A/\{x_1,\ldots,x_{i-1},x_{i+1},\ldots,x_m\}}(k_1, \ldots, k_{i-1}, k_{i+1}, \ldots, k_m) =$$

$$= \sum_{k_i \in \{0,1\}} \chi_{A/\{x_1,\ldots,x_m\}}(k_1, \ldots, k_m).$$

Inclusion.

The stochastic set A/X is *included* in the stochastic set B/X, written as $A/X \subset B/X$, if:

$$\chi_{A/X,B/X}(i_1, \ldots, i_m, j_1, \ldots, j_m) = 0$$

whenever there is at least a pair (i_k, j_k), $(1 \leq k \leq m)$, such that $j_k = 1$ and $i_k = 0$.

Compatibility.

Two stochastic sets A/X and B/X are *compatible* if $A/X \subset B/X$ and $B/X \subset A/X$.

Equality.

Two stochastic sets are equal, i.e., $A/X = B/X$, if $\chi_{A/X} = \chi_{B/X}$, which means:

$$\chi_{A/X}(i_1, \ldots, i_m) = \chi_{B/X}(i_1, \ldots, i_m),$$

for all values of $i_1, \ldots, i_m \in \{0,1\}$. Obviously,

$$A/\{x_1, \ldots, x_m\} = A/\{x_1\} \otimes \ldots \otimes A/\{x_m\}.$$

Independent entities.

The elements of universe X are independent with respect to property A if the following equality:

$$\chi_{A/\{x_1,\ldots,x_m\}}(i_1, \ldots, i_m) = \chi_{A/\{x_1\}}(i_1) \cdots \chi_{A/\{x_m\}}(i_m)$$

holds for all vectors:

$$(i_1, \ldots, i_m) \in \{0, 1\}^m.$$

Particular cases.

(a) A *fuzzy set* (Zadeh, (1965)) is defined by a membership function which can take on any real values from the unit interval $[0, 1]$. Instead of saying that an element x from the universe X belongs or does not belong to a set A, we assign to x a number $f_A(x), (0 \leq f_A(x) \leq 1)$, representing the degree of membership of x to A.

If $f_A : X \longrightarrow [0, 1]$ is a fuzzy set and the elements of the universe $X = \{x_1, \ldots, x_m\}$ are independent entities, then it generates the particular stochastic set:

$$\chi_{A/X}(k_1, \ldots, k_m) = \chi_{A/\{x_1, \ldots, x_m\}}(k_1, \ldots, k_m) =$$

$$= \varphi\{A/\{x_1\}, \ldots, A/\{x_m\}\}[k_1, \ldots, k_m] =$$

$$= \varphi\{A/\{x_1\}\}[k_1] \ldots \varphi\{A/\{x_m\}\}[k_m],$$

where:

$$\varphi\{A/\{x_i\}\}[1] = f_A(x_i),$$

$$\varphi\{A/\{x_i\}\}[0] = 1 - f_A(x_i),$$

for all values $(i = 1, \ldots, m)$.

Conversely, let A/X be a stochastic set with the membership probability distribution χ_A, and let $A/\{x_1\}, \ldots, A/\{x_m\}$ be its restrictions. The stochastic set A/X induces the fuzzy set on X whose membership function $f_A : X \longrightarrow [0, 1]$ is defined by:

$$f_A(x_i) = \chi_{A/\{x_i\}}(1), \quad (i = 1, \ldots, m).$$

Therefore, every stochastic set on X relative to a property (predicate) A induces a fuzzy set on X relative to A but the converse is true only if the elements of the universe X are independent with respect to A.

(b) A *crisp (Cantor) set* A is a particular stochastic set whose membership probability distribution χ_A is degenerate, i.e., there is only one

$m-$dimensional vector $(k_1, \ldots, k_m)$ such that $\chi_A(k_1, \ldots, k_m) = 1$. The corresponding crisp subset of X is $A = \{x_i; k_i = 1\}$. Conversely, to any crisp subset:

$$A = \{x_{i(1)}, \ldots, x_{i(r)}\} \subseteq X$$

it corresponds the stochastic set χ_A for which $\chi_A(k_1, \ldots, k_m)$ is equal to

$$\begin{cases} 1 & \text{if } k_{i(1)} = \ldots = k_{i(r)} = 1, k_i = 0, (i \neq i(1), \ldots, i(r)); \\ \\ 0 & \text{for all the other vectors } (k_1, \ldots, k_m). \end{cases}$$

Independent propositions.

Two propositions A, B are independent with respect to the universe X if:

$$\chi_{A/X,B/X} = \chi_{A/X}\,\chi_{B/X}.$$

Intersection.

If A/X and B/X are two stochastic sets then their *intersection* $A \cap B/X$ is the stochastic set having the membership probability distribution defined by:

$$\chi_{(A \cap B)/X}(i_1, \ldots, i_m) = \sum{}^{*}\chi_{A/X,B/X}(j_1, \ldots, j_m; k_1, \ldots, k_m),$$

where $\sum^{*}$ is taken in the following way: if $i_s = 1, (1 \leq s \leq m)$, then the sum contains all the terms for which $j_s = 1$ and $k_s = 1$; if $i_s = 0$, then the sum contains all the terms for which *either* $j_s = 1$ and $k_s = 0$ *or* $j_s = 0$ and $k_s = 1$, *or* $j_s = 0$ and $k_s = 0$. Obviously, if the two properties A and B are independent then, in the above sum, $\chi_{A/X,B/X}$ should be replaced by the product $\chi_{A/X}\,\chi_{B/X}$.

Remark: If A and B are independent properties (predicates) and the elements of the universe X are independent with respect to these two properties (predicates) then the intersection of the fuzzy sets f_A and f_B induced by the stochastic sets $\chi_{A/X}$ and $\chi_{B/X}$, respectively, is the fuzzy set:

$$f_{A \cap B}(x_i) = \chi_{(A \cap B)/\{x_i\}}(1) =$$

$$= \chi_{A/\{x_i\},B/\{x_i\}}(1; 1) = \chi_{A/\{x_i\}}(1)\,\chi_{B/\{x_i\}}(1) =$$

$$= f_A(x_i)\, f_B(x_i),$$

which is a generalization for fuzzy sets of the usual intersection between Cantor crisp sets.

Union.

If A/X and B/X are two stochastic sets then their *union* $A \cup B/X$ is the stochastic set having the membership probability distribution defined by:

$$\chi_{(A \cup B)/X}(i_1, \ldots, i_m) = \sum{}^{**} \chi_{A/X,B/X}(j_1, \ldots, j_m; k_1, \ldots, k_m)$$

where $\sum^{**}$ is taken in the following way: if $i_s = 0, (1 \leq s \leq m)$, then the sum contains all the terms for which $j_s = 0$ and $k_s = 0$; if $i_s = 1$, then the sum contains all the terms for which *either* $j_s = 1$ and $k_s = 0$ *or* $j_s = 0$ and $k_s = 1$, *or* $j_s = 1$ and $k_s = 1$. Obviously, if the two properties A and B are independent then, in the above sum, $\chi_{A/X,B/X}$ should be replaced by the product $\chi_{A/X}\, \chi_{B/X}$.

Remark: If A and B are independent properties (predicates) and the elements of the universe X are independent with respect to these two properties (predicates) then the intersection of the fuzzy sets f_A and f_B induced by the stochastic sets $\chi_{A/X}$ and $\chi_{B/X}$, respectively, is the fuzzy set:

$$f_{A \cup B}(x_i) = \chi_{(A \cup B)/\{x_i\}}(1) = \chi_{A/\{x_i\}}(1)\, \chi_{B/\{x_i\}}(0)+$$

$$+\chi_{A/\{x_i\}}(0)\, \chi_{B/\{x_i\}}(1) + \chi_{A/\{x_i\}}(1)\, \chi_{B/\{x_i\}}(1) =$$

$$= f_A(x_i)\,[1 - f_B(x_i)] + [1 - f_A(x_i)]\, f_B(x_i) + f_A(x_i)\, f_B(x_i) =$$

$$= f_A(x_i) + f_B(x_i) - f_A(x_i)\, f_B(x_i),$$

which is a generalization for fuzzy sets of the usual union between Cantor crisp sets.

Distributivity.

From the above definitions, if A, B, C are properties, then:

$$(A \cup (B \cap C))/X = ((A \cup B) \cap (A \cup C))/X;$$

$$(A \cap (B \cup C))/X = ((A \cap B) \cup (A \cap C))/X.$$

Complement.

If A/X is a stochastic set and $\chi_{A/X}$ its membership probability distribution, then its *complement* (or *negation*) with respect to property A is the stochastic set $\bar{A}/X$ whose membership probability distribution is:

$$\chi_{\bar{A}/X}(i_1, \ldots, i_m) = \chi_{A/X}(\bar{i}_1, \ldots, \bar{i}_m),$$

where $\bar{0} = 1$ and $\bar{1} = 0$.

Remark: If $f_A : X \longrightarrow [0,1]$ is a fuzzy set on X, then the membership function of the complementary fuzzy set is $f_{\bar{A}} = 1 - f_A$. If the elements of X are independent then f_A and $f_{\bar{A}}$ completely determine the membership probability distributions $\chi_{A/X}$ and $\chi_{\bar{A}/X}$ of the stochastic sets A/X and $\bar{A}/X$, respectively. As:

$$\chi_{A/\{x_i\}}(k) = \begin{cases} f_A(x_i), & \text{if } k = 1, \\ 1 - f_A(x_i), & \text{if } k = 0, \end{cases}$$

whereas:

$$\chi_{\bar{A}/\{x_i\}}(k) = \begin{cases} f_{\bar{A}}(x_i) = 1 - f_A(x_i), & \text{if } k = 1, \\ 1 - f_{\bar{A}}(x_i) = f_A(x_i), & \text{if } k = 0, \end{cases}$$

we have:

$$\chi_{\bar{A}/\{x_i\}}(k) = \chi_{A/\{x_i\}}(\bar{k}).$$

De Morgan's relations.

If A and B are two two-valued sentences or properties (predicates) referring to the elements of the same universe X, then we have:

Proposition 4.1: *For any two stochastic sets A/X and B/X we have:*

$$(\overline{A \cup B})/X = (\bar{A} \cap \bar{B})/X;$$

Proof: Using the definition of negation, union, and intersection, we have:

$$\chi_{(\overline{A \cup B})/X}(j_1, \ldots, j_m) =$$

$$= \chi_{(A \cup B)/X}(\bar{j}_1, \ldots, \bar{j}_m) =$$

$$= \sum{}^{**} \chi_{A/X,B/X}(k_1, \ldots, k_m, l_1, \ldots, l_m) =$$

$$= \sum{}^{*} \chi_{A/X,B/X}(\bar{k}_1, \ldots, \bar{k}_m, \bar{l}_1, \ldots, \bar{l}_m) =$$

$$= \sum{}^{*} \chi_{\bar{A}/X,\bar{B}/X}(k_1, \ldots, k_m, l_1, \ldots, l_m) =$$

$$= \chi_{(\bar{A} \cap \bar{B})/X}(j_1, \ldots, j_m),$$

where:

(a) $\sum^{**}$ is taken in the following way: if $\bar{j}_i = 0$, then the sum contains all the terms for which $k_i = 0$ and $l_i = 0$; if $\bar{j}_i = 1$, then the sum contains all the terms for which *either* $k_i = 1$ and $l_i = 0$ *or* $k_i = 0$ and $l_i = 1$, *or* $k_i = 1$ and $l_i = 1$.

(b) $\sum^{*}$ is taken in the following way: if $j_i = 1$, then the sum contains all the terms for which $k_i = 1$ and $l_i = 1$; if $j_i = 0$, then the sum contains all the terms for which *either* $k_i = 0$ and $l_i = 1$ *or* $k_i = 1$ and $l_i = 0$, *or* $k_i = 0$ and $l_i = 0$. $\quad \square$

Obviously, if the two propositions A and B are independent, then, in the above sum, $\chi_{A/X,B/X}$ is to be replaced by the product $\chi_{A/X} \chi_{B/X}$.

Proposition 4.2: *For any two stochastic sets A/X and B/X we have:*

$$(\overline{A \cap B})/X = (\overline{A} \cup \overline{B})/X.$$

Proof: Using the definition of negation, union, and intersection, we have:

$$\chi_{(\overline{A \cap B})/X}(j_1, \ldots, j_m) =$$

$$= \chi_{(A \cap B)/X}(\bar{j}_1, \ldots, \bar{j}_m) =$$

$$= \sum{}^{*} \chi_{A/X,B/X}(k_1, \ldots, k_m, l_1, \ldots, l_m) =$$

$$= \sum{}^{**} \chi_{A/X,B/X}(\bar{k}_1, \ldots, \bar{k}_m, \bar{l}_1, \ldots, \bar{l}_m) =$$

$$= \sum{}^{**} \chi_{\bar{A}/X,\bar{B}/X}(k_1, \ldots, k_m, l_1, \ldots, l_m) =$$

$$= \chi_{(\overline{A} \cup \overline{B})/X}(j_1, \ldots, j_m),$$

where:

(a) $\sum^*$ is taken in the following way: if $\bar{j}_i = 1$, then the sum contains all the terms for which $k_i = 1$ and $l_i = 1$; if $\bar{j}_i = 0$, then the sum contains all the terms for which *either* $k_i = 1$ and $l_i = 0$ *or* $k_i = 0$ and $l_i = 1$, *or* $k_i = 0$ and $l_i = 0$.

(b) $\sum^{**}$ is taken in the following way: if $j_i = 0$, then the sum contains all the terms for which $k_i = 0$ and $l_i = 0$; if $j_i = 1$, then the sum contains all the terms for which *either* $k_i = 0$ and $l_i = 1$ *or* $k_i = 1$ and $l_i = 0$, *or* $k_i = 1$ and $l_i = 1$. $\square$

Obviously, if the two propositions A and B are independent, then, in the above sum, $\chi_{A/X,B/X}$ is to be replaced by the product $\chi_{A/X}\,\chi_{B/X}$.

The law of double negation.

If A is a two-valued sentence or property (predicate) referring to the elements of the universe X, then we have:

Proposition 4.3: *For any stochastic set A/X we have:*

$$\overline{\overline{A}}/X = A/X.$$

Proof: Using the definition of negation twice, we get:

$$\chi_{\overline{\overline{A}}/X}(k_1,\ldots,k_m) = \chi_{\overline{A}/X}(\bar{k}_1,\ldots,\bar{k}_m) =$$

$$= \chi_{A/X}(\bar{\bar{k}}_1,\ldots,\bar{\bar{k}}_m) = \chi_{A/X}(k_1,\ldots,k_m). \quad \square$$

Conditional stochastic sets.

If the joint stochastic set $A/X \otimes B/Y$ is given and A/X and B/Y are its restrictions to universes X and Y, respectively, then denote by $A/X \mid B/Y$ the family of conditional stochastic sets relative to property A in universe X given property B in universe Y, defined by the conditional membership probability distributions:

$$\chi_{A/X|B/Y}(i_1,\ldots,i_m \mid j_1,\ldots,j_n) =$$

$$= \chi_{A/X, B/Y}(i_1, \ldots, i_m; j_1, \ldots, j_n)/\chi_{B/Y}(j_1, \ldots, j_n),$$

for those binary values of $j_1, \ldots, j_n$ for which the denominator is different from zero. In an abbreviated form, we can write the above equalities as:

$$\chi_{A/X, B/Y} = \chi_{A/X|B/Y} \, \chi_{B/Y}.$$

A similar definition may be given for $B/Y \mid A/X$, in which case:

$$\chi_{A/X, B/Y} = \chi_{B/Y|A/X} \, \chi_{A/X}.$$

Thus, if $m = n = 2$, for instance, $\chi_{A/X|B/Y}(1, 0 \mid 0, 1)$ is the probability (or credibility) that x_1 has *and* x_2 has not property A *if* y_1 has not *and* y_2 has property B.

Syllogisms.

(a) *Modus ponens.*

Given $B/Y \mid A/X$, if A/X then B/Y, where:

$$\chi_{B/Y}(j_1, \ldots, j_n) =$$

$$= \sum_{i_1, \ldots, i_m \in \{0,1\}} \chi_{B/Y|A/X}(j_1, \ldots, j_n \mid i_1, \ldots, i_m) \, \chi_{A/X}(i_1, \ldots, i_m).$$

The above equality may be abbreviated as:

$$\chi_{B/Y} = \chi_{B/Y|A/X} \odot \chi_{A/X}.$$

Let us notice also that if $\alpha \leq \chi_{B/Y|A/X} \leq \beta$, then $\alpha \leq \chi_{B/Y} \leq \beta$.

(b) *Modus tollens.*

Given $B/Y \mid A/X$, if B/Y then A/X, provided that the system of liniar equations:

$$\chi_{B/Y|A/X} \odot \chi_{A/X} = \chi_{B/Y}$$

may be solved with respect to the unknown probability (credibility) distribution $\chi_{A/X}$.

Belief and plausibility induced by a stochastic set.

Let A/X be a stochastic set with the membership probability (credibility) distribution $\chi_{A/X} : \{0,1\}^m \longrightarrow [0,1]$. The correspondence between the binary 0-1 vector $(i_1, \ldots, i_m)$ and the subset:

$$E = \{x_{i_j}; i_j = 1, 1 \leq j \leq m\} \subseteq X = \{x_1, \ldots, x_m\},$$

induces the equivalence between $\{0,1\}^m$ and the class of all subsets of X, namely $\mathcal{P}(X)$. The function $\chi_{A/X}$ induces a probability (credibility) distribution on $\mathcal{P}(X)$, denoted by χ_A, called basic probability assignment on $\mathcal{P}(X)$, i.e.,

$$\chi_A(E) = \chi_{A/X}(i_1, \ldots, i_m) \geq 0, \qquad \sum_{E \subseteq X} \chi_A(E) = 1.$$

Let $\mathcal{F}_A$ be the class of *focal subsets* of X with respect to χ_A, i.e.,

$$\mathcal{F}_A = \{E; E \subseteq X, \chi_A(E) > 0\} \subseteq \mathcal{P}(X).$$

The *belief* and *plausibility* induced by χ_A on $\mathcal{P}(X)$ are defined by:

$$Bel_A(E) = \sum_{F \subseteq E, F \neq \emptyset} \chi_A(F),$$

$$Pl_A(E) = \sum_{F \cap E \neq \emptyset} \chi_A(F),$$

for every nonempty $E \in \mathcal{P}(X)$. The standard definitions of the belief and plausibility functions assume that the basic probability assignment on $\mathcal{P}(X)$ has to be equal to zero for the empty subset of X. This is not the case here, as $\chi_A(\emptyset)$, which is $\chi_{A/X}(0, \ldots, 0)$ is not necessarily equal to zero. In our case, for each nonempty $E \subset X$ we have:

$$\chi_A(\emptyset) + Bel_A(E) + Pl_A(X - E) = 1.$$

A positive value of $\chi_{A/X}(0, \ldots, 0)$ corresponds to the possible case when no element of the universe X has the property described by sentence A.

Remark: As in the standard case, if the elements of $\mathcal{F}_A$ are nested, i.e., $E_1 \supseteq E_2 \supseteq \ldots$, then the belief and plausibility induced by χ_A on $\mathcal{P}(X)$ are called *necessity* and *possibility*, respectively.

Measures of uncertainty and interdependence.

As the stochastic logic deals with membership probability (credibility) distributions, the classical measures of uncertainty may be used. Thus, denoting by $H(A/X)$ the Shannon entropy of the probability (credibility) distribution $\chi_{A/X}$, it gives the amount of uncertainty contained by this probability distribution. We have:

$$H(A/X) = H(A/\{x_1, \ldots, x_m\}) =$$

$$= - \sum_{(k_1, \ldots, k_m) \in \{0,1\}^m} \chi_{A/X}(k_1, \ldots, k_m) \, \log \chi_{A/X}(k_1, \ldots, k_m).$$

Watanabe's measure of connection or interdependence between the restrictions A/X and B/Y of the joint stochastic set $A/X \otimes B/Y$ is:

$$W(A/X \otimes B/Y; A/X, B/Y) = H(A/X) + H(B/Y) - H(A/X \otimes B/Y).$$

In particular, the amount of interdependence among the entities of the universe $X = \{x_1, \ldots, x_m\}$ with respect to property A is:

$$W(A/\{x_1, \ldots, x_m\}; A/\{x_1\}, \ldots, A/\{x_m\}) =$$

$$= \sum_{i=1}^{m} H(A/\{x_i\}) - H(A/\{x_1, \ldots, x_m\}),$$

where:

$$H(A/\{x_i\}) = -\chi_{A/\{x_i\}}(0) \, \log \chi_{A/\{x_i\}}(0) - \chi_{A/\{x_i\}}(1) \, \log \chi_{A/\{x_i\}}(1).$$

The entropic distance between the stochastic sets A/X and B/Y is:

$$d(A/X, B/Y) = H(A/X \otimes B/Y) - W(A/X \otimes B/Y; A/X, B/Y).$$

Remark: For a fuzzy set A with the membership function:

$$f_A : X \longrightarrow [0, 1], \quad X = \{x_1, \ldots, x_m\},$$

the corresponding stochastic set with independent elements is:

$$\chi_{A/X}(k_1, \ldots, k_m) = \varphi\{A/\{x_1\}, \ldots, A/\{x_m\}\}[k_1, \ldots, k_m] =$$

$$= \varphi\{A/\{x_1\}\}[k_1] \ldots \varphi\{A/\{x_m\}\}[k_m],$$

where:

$$\varphi\{A/\{x_i\}\}[1] = f_A(x_i), \quad \varphi\{A/\{x_i\}\}[0] = 1 - f_A(x_i),$$

for all values $(i = 1, \ldots, m)$, where each k_i may take on only two logical values, namely, $k_i \in \{0, 1\}$. If we calculate the entropy of the joint probability distribution:

$$\chi_{A/X}(k_1, \ldots, k_m) = \varphi\{A/\{x_1\}\}[k_1] \ldots \varphi\{A/\{x_m\}\}[k_m],$$

we get the particular expression:

$$H(A/X) = - \sum_{(k_1, \ldots, k_m) \in \{0,1\}^m} \chi_{A/X}(k_1, \ldots, k_m) \log \chi_{A/X}(k_1, \ldots, k_m) =$$

$$= \sum_{i=1}^{m} [-\varphi\{A/\{x_i\}\}[1] \log \varphi\{A/\{x_i\}\}[1] - \varphi\{A/\{x_i\}\}[0] \log \varphi\{A/\{x_i\}\}[0]] =$$

$$= \sum_{i=1}^{m} [-f_A(x_i) \log f_A(x_i) - (1 - f_A(x_i)) \log(1 - f_A(x_i))].$$

The last expression has been called the entropy of the fuzzy set and was introduced by A. De Luca and S. Termini (1972) and called "nonprobabilistic entropy". We have seen that in fact it is just Shannon's entropy of the joint probability distribution:

$$\chi_{A/X}(k_1, \ldots, k_m)$$

when the elements of the universe $X = \{x_1, \ldots, x_m\}$ are independent.

It has been repeatedly mentioned in literature that the membership function f_A that defines a fuzzy set is not a probability distribution on the set X and that the fuzzy set theory has nothing to do with probability theory. Indeed, f_A is not a probability distribution on A but, nevertheless, it defines a probability distribution, namely,

$$f_A(x_i), \quad 1 - f_A(x_i),$$

for each element $x_i \in X$. If the elements of X are independent, the product of this family of probability distributions defines a joint probability distribution $\chi_{A/X}(k_1, \ldots, k_m)$ on $\{0, 1\}^m$.

Paradoxes.

It is well-known that self-referential paradoxes appear in classic logic when there is a set for which a proposition A is both true and false. There is no such paradox in stochastic logic because each stochastic set is itself self-contradictory. In particular, the extreme events "All the entities of the universe $X = \{x_1, \ldots, x_m\}$ have the property A," and "No element of X has the property A," may be possible if:

$$\chi_{A/\{x_1,\ldots,x_m\}}(1,\ldots,1) > 0, \quad \text{and} \quad \chi_{A/\{x_1,\ldots,x_m\}}(0,\ldots,0) > 0.$$

Generalizations.

1.) The above considerations may be generalized in a straightforward way to the case of a multi-valued stochastic logic. In such a case, the membership probability (credibility) distribution should be replaced by:

$$\chi_{A/X} : \{0_A, 1_A, 2_A, \ldots, M_A\}^m \longrightarrow [0,1],$$

where $0_A, 1_A, 2_A, \ldots, M_A$ are the possible values of sentence or property (predicate) A. In such a case:

$$\chi_{A/X}(k_1, \ldots, k_m), \quad k_i \in \{0_A, 1_A, 2_A, \ldots, M_A\},$$

is the probability that the element x_1 has the logical value k_1 of sentence or property (predicate) A, and,..., and the element x_m has the value k_m of sentence or property (predicate) A. In such a multi-valued case, the negation of a logical value k_i is a set of values, namely:

$$\bar{k}_i = \{0_A, 1_A, 2_A, \ldots, M_A\} - \{k_i\}.$$

2.) The above formalism may be extended to the case when a time instant, different or not, is assigned to each property, and we deal with the joint membership probability distribution:

$$\chi_{(A/X,t_1),(B/Y,t_2),\ldots}(i_1, \ldots, i_m, j_1, \ldots, j_n, \ldots).$$

4.3. Examples

1) *Numerical example.*

Using the notations from the previous section, let us take only two entities $X = \{x_1, x_2\}$, and two properties A, B. Given the stochastic set:

$$\chi_{A/X,B/X}(i_1, i_2, j_1, j_2) =$$

$$= \varphi\{A/\{x_1\}, A/\{x_2\}, B/\{x_1\}, B/\{x_2\}\}[i_1, i_2, j_1, j_2]$$

defined by the following tableaus:

$\chi_{A/X,B/X}$	0.05	0.06	0.20	0.10	0.02	0.00	0.09	0.10
$B/\{x_2\}$	0	1	0	1	0	1	0	1
$B/\{x_1\}$	0	0	1	1	0	0	1	1
$A/\{x_2\}$	0	0	0	0	1	1	1	1
$A/\{x_1\}$	0	0	0	0	0	0	0	0

$\chi_{A/X,B/X}$	0.20	0.05	0.00	0.06	0.04	0.01	0.02	0.00
$B/\{x_2\}$	0	1	0	1	0	1	0	1
$B/\{x_1\}$	0	0	1	1	0	0	1	1
$A/\{x_2\}$	0	0	0	0	1	1	1	1
$A/\{x_1\}$	1	1	1	1	1	1	1	1

we obtain from them the values of:

$$\chi_{A/X}(i_1, i_2) = \varphi\{A/\{x_1\}, A/\{x_2\}\}[i_1, i_2],$$

$$\chi_{\bar{A}/X}(i_1, i_2) = \varphi\{\bar{A}/\{x_1\}, \bar{A}/\{x_2\}\}[i_1, i_2],$$

$$\chi_{(A\cap B)/X}(k_1, k_2) = \varphi\{(A\cap B)/\{x_1\}, (A\cap B)/\{x_2\}\}[k_1, k_2],$$

$$\chi_{(A\cup B)/X}(k_1, k_2) = \varphi\{(A\cup B)/\{x_1\}, (A\cup B)/\{x_2\}\}[k_1, k_2],$$

respectively, namely:

$\chi_{A/X}$	0.41	0.21	0.31	0.07
$A/\{x_2\}$	0	1	0	1
$A/\{x_1\}$	0	0	1	1

$\lambda_{\bar{A}/X}$	0.07	0.31	0.21	0.41
$\bar{A}/\{x_2\}$	0	1	0	1
$\bar{A}/\{x_1\}$	0	0	1	1

$\lambda_{(A\cap B)/X}$	0.81	0.11	0.08	0.00
$(A\cap B)/\{x_2\}$	0	1	0	1
$(A\cap B)/\{x_1\}$	0	0	1	1

$\lambda_{(A\cup B)/X}$	0.05	0.08	0.40	0.47
$(A\cup B)/\{x_2\}$	0	1	0	1
$(A\cup B)/\{x_1\}$	0	0	1	1

2) *Stochastic sets induced by evidence.*

In an Agatha Christie's detective story (Christie, (1984a)), Lord Edgware is murdered. The main characters are:

j = Jane Wilkinson, an actress, the wife of Lord Edgware;

c = Carlotta Adams, an actress who can perfectly impersonate j;

b = Bryan Martin, an actor who loves j and is a friend of c;

r = Captain Ronald Marsh, the nephew of Lord Edgware, in great need of money;

s = Miss Carroll, Lord Edgware's secretary;

d = Jenny Driver, the friend of c;

g = Geraldine Marsh, Lord Edgware's daughter, who hated her father and is very fond of r;

a = Duchess of Merton, an aristocratic old lady who wants to do anything possible in order to prevent the marriage of her religious son with j;

h = Lord Edgware's handsome butler.

The case is investigated by the detective Hercule Poirot. Let X_i be the universe of suspects at the i-th step of the investigation. Let M_i be the property "having had a motive to kill Lord Edgware according to what is known at the i-th step of the investigation" and O_i the property "having had an opportunity to kill Lord Edgware according to what is known at the i-th step of the investigation." In his analysis, at each step of his investigation,

Hercule Poirot applies Laplace's cautious Principle of Insufficient Reason, according to which if there is no reason to discriminate between the possible outcomes then the best strategy is to take them as being equally likely. The entities belonging to the intersection $M_i \cap O_i/X_i$, abbreviated as P_i/X_i, are the potential murderers of Lord Edgware at the i-th step of the investigation. Let us follow Hercule Poirot's steps in investigating the case:

Step 1: Evidence #1 (Lord Edgware dies). $\Longrightarrow$

$$X_1 = \{j\};$$

$$\chi_{M_1/X_1,O_1/X_1}(1;1) = 1,$$

$$\chi_{P_1/X_1}(1) = 1.$$

Step 2: Evidence #1 and evidence #2 (Jane attended the dinner party at Sir Montagu Corner at the time when Lord Edgware was murdered). $\Longrightarrow$

$$X_2 = \{c\};$$

$$\chi_{M_2/X_2,O_2/X_2}(1;1) = 1,$$

$$\chi_{P_2/X_2}(1) = 1.$$

Step 3: Evidence #1, evidence #2, and evidence #3 (Carlotta dies and the letter sent to her sister incriminates Ronald). $\Longrightarrow$

$$X_3 = \{c,r\};$$

$$\chi_{M_3/X_3,O_3/X_3}(1,1;1,0) = 1/2,$$

$$\chi_{M_3/X_3,O_3/X_3}(1,1;1,1) = 1/2,$$

$$\chi_{P_3/X_3}(1,0) = 1/2,$$

$$\chi_{P_3/X_3}(1,1) = 1/2.$$

Step 4: Evidence #1, evidence #2, and evidence #4 (Poirot discovers that a page was ingeniously removed from Carlotta's letter which now may incriminate any men not only Ronald). $\Longrightarrow$

$$X_4 = \{c,r,b,h\};$$

$$\chi_{M_4/X_4,O_4/X_4}(1,1,1,0;1,0,0,0) = 1/4,$$

$$\chi_{M_4/X_4,O_4/X_4}(1,1,1,0;1,1,0,0) = 1/4,$$

$$\chi_{M_4/X_4,O_4/X_4}(1,1,1,0;1,0,1,0) = 1/4,$$

$$\chi_{M_4/X_4,O_4/X_4}(1,1,1,0;1,0,0,1) = 1/4;$$

$$\chi_{P_4/X_4}(1,0,0,0) = 1/2,$$

$$\chi_{P_4/X_4}(1,1,0,0) = 1/4,$$

$$\chi_{P_4/X_4}(1,0,1,0) = 1/4.$$

Step 5: Evidence #1, evidence #2, and evidence #4 (Poirot discovers that in Carlotta's letter the word *he* should be read *she* in which case the letter incriminates women not men). $\Longrightarrow$

$$X_5 = \{c,a,d,g,s\};$$

$$\chi_{M_5/X_5,O_5/X_5}(1,1,0,1,0;1,0,0,0,0) = 1/5,$$

$$\chi_{M_5/X_5,O_5/X_5}(1,1,0,1,0;1,1,0,0,0) = 1/5,$$

$$\chi_{M_5/X_5,O_5/X_5}(1,1,0,1,0;1,0,1,0,0) = 1/5,$$

$$\chi_{M_5/X_5,O_5/X_5}(1,1,0,1,0;1,0,0,1,0) = 1/5,$$

$$\chi_{M_5/X_5,O_5/X_5}(1,1,0,1,0;1,0,0,0,1) = 1/5;$$

$$\chi_{P_5/X_5}(1,0,0,0,0) = 3/5, \quad \chi_{P_5/X_5}(1,1,0,0,0) = 1/5,$$

$$\chi_{P_5/X_5}(1,0,0,1,0) = 1/5.$$

Step 6: Evidence #1, evidence #5, and evidence #6 (Jane's humiliating gaffe about the judgment of Paris at Widburns' luncheon party at Claridge's distroys evidence #2). $\Longrightarrow$

$$X_6 = \{j\};$$

$$\chi_{M_6/X_6,O_6/X_6}(1;1) = 1; \chi_{P_6/X_6}(1) = 1.$$

And Jane Wilkinson is arrested. She confesses murdering Lord Edgware.

4.4. Markov chains

Markov chains have originated in statistical mechanics and have many applications in mathematical modelling. They may be viewed, however, as a special case of relative logic. Let p represent "the state of a system" and

Q_i "the time instant i". Therefore, p/Q_i represents "the state of a system at time i". In such a case the set of logical values V will represent the possible states of the system. The system can be in only one state at each time i but there are several such possible states. In order to simplify the writing, denote the probability that the state of the system is k_{i+1} at time $i+1$ if its state was k_i at time i by:

$$\pi_{i,i+1}(k_{i+1} \mid k_i) = \varphi\{p/Q_{i+1} \mid p/Q_i\}[k_{i+1} \mid k_i].$$

The evolution of the system is a Markov chain if the joint probability distribution satisfies the equality:

$$\varphi\{p/Q_1, \ldots, p/Q_n\}[k_1, \ldots k_n] =$$
$$= \varphi\{p/Q_1\}[k_1] \, \varphi\{p/Q_2 \mid p/Q_1\}[k_2 \mid k_1] \ldots \varphi\{p/Q_n \mid p/Q_{n-1}\}[k_n \mid k_{n-1}] =$$
$$= \pi_1(k_1) \, \pi_{1,2}(k_2 \mid k_1) \ldots \pi_{n-1,n}(k_n \mid k_{n-1}),$$

where:

$$\pi_i(k_i) = \varphi\{p/Q_i\}[k_i].$$

The probability of having the system in state k_n at time n is:

$$\varphi\{p/Q_n\}[k_n] = \sum_{k_1,\ldots,k_{n-1}} \pi_1(k_1) \prod_{i=1}^{n-1} \pi_{i,i+1}(k_{i+1} \mid k_i).$$

The amount of uncertainty on the states of the system at the time instants $1, \ldots, n$ is:

$$H(p/Q_1 \otimes \ldots \otimes p/Q_n) = H(p/Q_1) + \sum_{i=1}^{n-1} H(p/Q_{i+1} \mid p/Q_i),$$

where:

$$H(p/Q_1) = -\sum_{k_1} \pi_1(k_1) \, \log \pi_1(k_1),$$

$$H(p/Q_{i+1} \mid p/Q_i) = -\sum_{k_i,k_{i+1}} \pi_i(k_i) \, \pi_{i,i+1}(k_{i+1} \mid k_i) \, \log \pi_{i,i+1}(k_{i+1} \mid k_i).$$

The Markov chain is stationary if the transition stochastic matrix does not depend on time, i.e., if:

$$\varphi\{p/Q_{i+1} \mid p/Q_i\}[k' \mid k] = \pi[k' \mid k],$$

for all time instants i and all states $k \in V, k' \in V$.

Chapter 5. APPLICATIONS

5.1. Pattern-recognition

1. Generalities. The capacity of performing pattern-recognition is one of the main qualities of the human intellect. The mathematical analysis of pattern-recognition is a central topic in artificial intelligence. Briefly, we have a set of entities defined, as complex or molecular sentences, in terms of possible logical values of some characteristics, described by simple or atomic sentences. Between some values of some characteristics we can have incompatibility relations. There is a prior probability distribution on the set of possible entities. The problem is to see what characteristics to examine, and in what order, for identifying the corresponding entity as fast as possible. To give an example, the entities could be some diseases (say kidney diseases) and the characteristics are symptoms as reveald by corresponding tests. If a new patient comes suffering from one of these diseases we want to see which tests have to be performed, and in which order, for getting the symptoms and eventually identifying the specific disease. More often than not, all tests are performed simultaneously and the diagnosis is made after looking at the results of all these tests. The problem is that the results of the tests are not independent and the characteristics are not equally important. We want to determine which is the most important test to start with and, depending on its result, which is the next test to be performed, and again, depending on its result, which is the next test to be performed, etc., until we can identify the actual disease. The mathematical analysis shows that few symptoms must be known, certainly nor all of them, in order to identify the specific disease. The solution of the pattern-recognition process is formulated in terms of a tree, each decision node of it telling which characteristic to check at that stage, essentially depending on the results obtained for the characteristics examined thus far. For each case we will follow the indications given in only one path of this tree and, at the end of this path, we get the corresponding entity. The prior probability distribution is given by the relative frequencies of the entities, based on past experience. If nothing is known and we have no reason to assume that they have different occurrence frequencies, we take the entities as being priorly equally likely. But even when the entities have equal prior probabilities or credibilities, the logical structure of the entities defined in terms of the logical values of the characteristics induces a certain hierarchy of the characteristics with respect to their relevance for the pattern-recognition of the corresponding entities.

The fact that the characteristics are not equally important is a main fact in practically all real life pattern-recognition cases. In almost all languages there is a popular proverb saying something like "Repetition is the mother of learning," and the educators are always expecting the pupils to practice as much as possible. But why do we learn to perform faster and better after many repetitions? The answer is that by repeating, the human intellect is able to realize first that some operations are useless and some are useful. Eventually, intelligent subjects realize that even among the useful operations, some are more useful than others and, by repetitions they succeed in focusing on the most relevant operations and in finding out the optimum order of performing the optimum operations in their interdependent order.

Remarkably enough, even if the characteristics are defined by atomic sentences taking on only two values (yes-no or present-absent) like the true-false values from the classic logic, the decision variable involved in any pattern-recognition problem, (i.e., "Which entity does occur?" or "Which entity to choose?") has more possible values (e.g., "The first entity does occur," ... "The m-th entity does occur," or "I choose the first entity," ... "I choose the m-th entity," respectively), normally requiring a multi-valued logic.

Pattern-recognition is an ideal topic for showing that classic two-valued or multi-valued logic is necessary but not sufficient for getting the optimum strategy of acting. The probability distribution induced both by the prior frequencies of the entities and by the logical structure of the entities in terms of the corresponding characteristics, together with the probabilistic measures of uncertainty, are both necessary for obtaining the optimum pattern-recognition strategy. At each stage, we are selecting the characteristic whose examination would minimize the remaining amount of uncertainty on the possible entities remained in competition. On each branch of the decision tree, we continue doing this until the entire uncertainty is eliminated. The final pattern-recognition tree includes all possibilities but when we apply it to a particular case, we are following only one path, depending on the specific results obtained for the characteristics examined step by step.

Let us consider n characteristics described by the simple (or atomic) sentences $p_1, \ldots, p_n$. Let V_j be the set of possible values of sentence p_j. Let us assume that there are m entities described by complex (or molecular) sentences $P_1, \ldots, P_m$. To each entity [defined by the molecular sentence] P_i there corresponds a distinct set A_i of vectors from the product set:

$$V_1 \times V_2 \times \ldots \times V_n,$$

such that the entity defined by P_i occurs (or is chosen) if the n-dimensional vector of values of the characteristics $p_1, \ldots, p_n$ belongs to the set A_i, i.e.,

$$(k_1, \ldots, k_n) \in A_i \subset \prod_{j=1}^{n} V_j.$$

The entity P_i does not occur (or is not chosen) if:

$$(k_1, \ldots, k_n) \notin A_i.$$

Therefore, the set of possible values of the entities $P_1, \ldots, P_m$ is $\tilde{V} = \{0, 1\}$, where 1 means "occurs", or "is chosen", and 0 means "does not occur", or "is not chosen".

The sets $A_1, \ldots, A_m$ are disjoint, which means that, in general, there are several vectors $(k_1, \ldots, k_n)$ corresponding to the same entity but no such vector could belong to two different sets A_i and A_ℓ. Let us notice that if two sets A_i and A_ℓ, corresponding to the entities P_i and P_ℓ, respectively, are not disjoint and there is at least one vector $(k_1, \ldots, k_n)$ belonging to both sets A_i and A_ℓ, we can always replace the two entities P_i and P_ℓ by three entities, say P_i', P_ℓ', and $P_{i,\ell}$ corresponding to the disjoint sets of vectors $(k_1, \ldots, k_n)$:

$$A_i - A_\ell, \qquad A_\ell - A_i, \qquad A_i \cap A_\ell,$$

respectively.

Let assume that we know the prior probability (credibility) distribution:

$$\varphi_0\{p_1, \ldots, p_n\}[k_1, \ldots, k_n] \geq 0, \quad (k_1, \ldots, k_n) \in \bigcup_{i=1}^{m} A_i.$$

We have:

$$\sum_{i=1}^{m} \sum_{(k_1, \ldots, k_n) \in A_i} \varphi_0\{p_1, \ldots, p_n\}[k_1, \ldots, k_n] = 1.$$

This probability distribution is generally based on past experience and past observations. Thus, if the entities are diseases and the characteristics are symptoms or results of some medical tests, the prior probability distribution just mentioned is obtained from statistics giving the frequency of the different vectors of values of the characteristics corresponding to the different diseases under investigation. If, however, we have no previous such data or if we have no reason to discriminate between the entities, as far as their occurrence frequency is concerned, we simply take them to be equally likely.

Let us calculate the initial probability that the entity P_i occurs, namely,

$$\varphi_0\{P_i\}[1] = \sum_{(k_1,\ldots,k_n)\in A_i} \varphi_0\{p_1,\ldots,p_n\}[k_1,\ldots,k_n],$$

for all values of i, $(i = 1,\ldots,m)$, in which case:

$$\varphi_0\{P_i\}[0] = 1 - \varphi_0\{P_i\}[1]$$

is the initial probability that entity P_i does not occur. These initial probabilities of the entities (i.e., possible diseases, or alternatives) are exclusively based on the prior distribution of the possible values of the characteristics (i.e., symptoms, types of evidence) obtained from past experience or reflecting our credibility, before relying on the result of any test that could reveal the actual value of any of the characteristics for the particular case (i.e., patient, new case under investigation) we are analysing.

Finally, let us introduce the *decision variable* defined by the sentence P which refers to: "The entity which does occur, or is chosen/selected." The set of its possible values is:

$$V^* = \{\text{"Entity } P_1 \text{ occurs."};\ldots;\text{"Entity } P_m \text{ occurs."}\},$$

or, equivalently,

$$V^* = \{\text{"Entity } P_1 \text{ is chosen."};\ldots;\text{"Entity } P_m \text{ is chosen."}\}.$$

The prior probabilities (credibilities) of these possible values are:

$$\{\varphi_0\{P_i\}[1],\ (i = 1,\ldots,m)\}.$$

We have:

$$\sum_{i=1}^{m} \varphi_0\{P_i\}[1] = 1.$$

We want to determine now which characteristic should be verified first. The best characteristic is the one whose known value most decreases the amount of uncertainty on the available entities. Initially, the amount of uncertainty on the possible values of the decision variable P, or equivalently, the amount of uncertainty on the set of available entities, is:

$$H_0(P) = -\sum_{i=1}^{m} \varphi_0\{P_i\}[1] \, \log \, \varphi_0\{P_i\}[1].$$

Let us focus on an arbitrary characteristic p_j. We calculate the probability that characteristic p_j has the value $k \in V_j$, under the circumstance A_i, i.e., for all vectors $(k_1, \ldots, k_n)$ making up the set A_i, corresponding to entity P_i, namely,

$$\varphi_0\{p_j/A_i\}[k] = \frac{\sum_{\{(k_1,\ldots,k_n)\in A_i, k_j=k\}} \varphi_0\{p_1,\ldots,p_n\}[k_1,\ldots,k_n]}{\sum_{(k_1,\ldots,k_n)\in A_i} \varphi_0\{p_1,\ldots,p_n\}[k_1,\ldots,k_n]},$$

for every $k \in V_j$. Now, we can calculate the probability that characteristic p_j has the value $k \in V_j$, taking all entities $P_1, \ldots, P_m$, or, equivalently, all sets $A_1, \ldots, A_m$, into account, namely,

$$\varphi_0\{p_j\}[k] = \sum_{i=1}^{m} \varphi_0\{P_i\}[1] \, \varphi_0\{p_j/A_i\}[k],$$

for every $k \in V_j$. According to Bayes' formula, the probability that the arbitrary entity P_i occurs, or is chosen, if we know that the value of p_j is k is:

$$\varphi_1\{P_i \mid p_j\}[1 \mid k] = \frac{\varphi_0\{P_i\}[1] \, \varphi_0\{p_j/A_i\}[k]}{\varphi_0\{p_j\}[k]},$$

for every $k \in V_j$ and $i = 1, \ldots, m$.

The conditional entropy of the decision variable P, taken all the values of characteristic p_j into account, is:

$$H_1(P \mid p_j) =$$

$$= \sum_{k\in V_j} \varphi_0\{p_j\}[k] \left(-\sum_{i=1}^{m} \varphi_1\{P_i \mid p_j\}[1 \mid k] \, \log \, \varphi_1\{P_i \mid p_j\}[1 \mid k] \right).$$

The number $H_1(P \mid p_j)$ shows the amount of uncertainty remaining on the possible values of the decision variable P if we have checked (i.e., observed, determined, measured, assumed) the value of characteristic p_j. Obviously, the most important characteristic is the one for which this amount of remaining uncertainty is minimum, i.e., we select the best characteristic p_ℓ for which:

$$H_1(P \mid p_\ell) = \min_{1 \leq j \leq n} H_1(P \mid p_j).$$

Assuming now that we know the particular value of characteristic p_ℓ, we repeat the same analysis for finding the next characteristic to be checked. The analysis is done for all possible values of p_ℓ. Thus, let us assume that checking (i.e., observing, determining, measuring) the optimum characteristic p_ℓ we have found that its value is $s \in V_\ell$. We are repeating the computations mentioned above but now, under the circumstance:

$$Q : \text{``}\varphi_1\{p_\ell\}[s] = 1,\text{''}$$

which, using words, means:

Q : "Assume that characteristic p_ℓ has, with certainty, the value s."

Let us define the set $B_i = A_i/Q \subseteq A_i$, containing only those vectors $(k_1, \ldots, k_n)$ of the set A_i that have the value s in the ℓ-th component k_ℓ, namely,

$$B_i = \{(k_1, \ldots, k_n) : (k_1, \ldots, k_n) \in A_i, k_\ell = s\}.$$

We do this for all values $(i = 1, \ldots, m)$. The subset B_i may coincide with A_i, or become the empty set $\emptyset$ but, generally, it is a subset in between $\emptyset$ and A_i. Obviously, if $B_i = \emptyset$, then it and the corresponding entity P_i are ignored. We calculate the probability of the entities under circumstance Q, i.e.,

$$\varphi_1\{P_i/Q\}[1] = \frac{\sum_{(k_1,\ldots,k_n) \in B_i} \varphi_0\{p_1, \ldots, p_n\}[k_1, \ldots, k_n]}{\sum_{(k_1,\ldots,k_n) \in B_1 \cup \ldots \cup B_m} \varphi_0\{p_1, \ldots, p_n\}[k_1, \ldots, k_n]},$$

for every $(i = 1, \ldots, m)$. The probability that characteristic p_j, $(j \neq \ell)$, has the value $k \in V_j$, under circumstance B_i, i.e., for the vectors $(k_1, \ldots, k_n)$ of the set B_i, is:

$$\varphi_1\{p_j/B_i\}[k] = \frac{\sum_{\{(k_1,\ldots,k_n) \in B_i, k_j = k\}} \varphi_0\{p_1, \ldots, p_n\}[k_1, \ldots, k_n]}{\sum_{(k_1,\ldots,k_n) \in B_i} \varphi_0\{p_1, \ldots, p_n\}[k_1, \ldots, k_n]},$$

for every $k \in V_j$. We calculate the probability that characteristic p_j has the value $k \in V_j$, taking those entities $P_1, \ldots, P_m$ that are still available, or, equivalently, all those subsets $B_1, \ldots, B_m$ that are nonempty, into account, namely,

$$\varphi_1\{p_j/Q\}[k] = \sum_{i=1}^{m} \varphi_1\{P_i/Q\}[1]\,\varphi_1\{p_j/B_i\}[k],$$

for every $k \in V_j$, $(j \neq \ell)$. According to Bayes' formula, the probability that the arbitrary entity P_i occurs, or is chosen, under circumstance Q, if we know that the value of p_j is k is:

$$\varphi_2\{P_i/Q \mid p_j\}[1 \mid k] = \frac{\varphi_1\{P_i/Q\}[1]\,\varphi_1\{p_j/B_i\}[k]}{\varphi_1\{p_j/Q\}[k]},$$

for every $k \in V_j$, $j \neq \ell$, and $i = 1, \ldots, m$.

The conditional entropy of the decision variable P, under circumstance Q, taking all possible values of characteristic p_j, $(j \neq \ell)$, into account is:

$$H_2(P/Q \mid p_j) =$$

$$= \sum_{k \in V_j} \varphi_1\{p_j/Q\}[k] \left(-\sum_{i=1}^{m} \varphi_2\{P_i/Q \mid p_j\}[1 \mid k] \log \varphi_2\{P_i/Q \mid p_j\}[1 \mid k] \right).$$

The number $H_2(P/Q \mid p_j)$ shows the amount of uncertainty still remaining on the possible values of the decision variable P, i.e., on the set of the entities still available, if we had obtained the value s for charactersitic p_ℓ, at the first step of the investigation, and now, we have checked (i.e., observed, determined, measured) the characteristic p_j, $(j \neq \ell)$, at the second step. Obviously, the most important characteristic, under circumstance Q, is the one for which this amount of remaining uncertainty is minimum, i.e., we select the characteristic p_r for which:

$$H_2(P/Q \mid p_r) = \min_{1 \leq j \leq n, j \neq \ell} H_2(P/Q \mid p_j).$$

Therefore, we start the pattern-recognition algorithm by checking (i.e., observing, determining, measuring) the characteristic p_ℓ. If its value proves to be $s \in V_\ell$, then the next characteristic to be checked is p_r. If the value of

p_r is found to be $t \in V_r$, we repeat the procedure described above, under the new circumstance:

$$Q : \text{``}\varphi_1\{p_\ell\}[s] = 1, \ \varphi_2\{p_r\}[t] = 1,\text{''}$$

in order to find which characteristic:

$$p_j, \quad 1 \leq j \leq n, \quad j \neq \ell, j \neq r,$$

should be the next one to be checked. We continue our path until the remaining uncertainty on the possible values of the decision variable P is equal to zero, in which case, given the values $s, t, \ldots$ of the most relevant characteristics $p_\ell, p_r, \ldots$, respectively, only one entity remains possible.

This is how one possible path of the pattern-recognition algorithm is obtained. The same analysis has to be performed going back to the first step of the algorithm, where the characteristic p_ℓ was to be checked, and assume that another value $s' \in V_\ell$ of this characteristic has been obtained. We follow the same strategy described above for finding the next characteristic to be checked under the new circumstance:

$$Q : \text{``}\varphi_1\{p_\ell\}[s'] = 1,\text{''}$$

induced by assuming a different value $s' \in V_\ell$ for p_ℓ. This new characteristic will not necessarily be p_r as it happened when we had assumed that the value of characteristic p_ℓ had been s. We continue the analysis until at the end of this new path of the algorithm a final unique entity is identified as well. And so on. The branches of the final pattern-recognition algorithm give all possible paths corresponding to the different consecutive values of the characteristics (not always the same) that have to be checked at the different steps (i.e., nodes) of the algorithm. These paths have different lengths (i.e., number of steps to be performed along them). Although the algorithm contains several paths ending into the corresponding final entities, when the pattern-recognition strategy is applied to a specific case, we are going to follow only one of these paths. Going along it, we are told, at each step, what characteristic to check, essentially depending on what values we have obtained thus far for the characteristics already checked at the previous successive steps.

2. *Example.* In what follows we will assume that $n = 4$ and each characteristic can take on only one of two possible values, namely, 1 for "present", and 0 for "absent". Therefore, we assume in our example that the set of logical values is binary and the same for all four characteristics, p_1, p_2, p_3, and p_4, i.e., $V_j = V = \{0, 1\}$. Assume also that there are five entities P_1, P_2, P_3, P_4, and P_5, corresponding to the sets A_1, A_2, A_3, A_4, and A_5 given below:

A_i	k_1	k_2	k_3	k_4
A_1	1	1	0	1
	0	1	1	1
	0	1	0	1
A_2	0	1	0	0
A_3	0	0	1	1
A_4	1	0	0	1
A_5	0	0	0	1

Assume that the following prior probability distribution is given:

$$\varphi_0\{p_1, p_2, p_3, p_4\}[1, 1, 0, 1] = 1/15,$$

$$\varphi_0\{p_1, p_2, p_3, p_4\}[0, 1, 1, 1] = 1/15,$$

$$\varphi_0\{p_1, p_2, p_3, p_4\}[0, 1, 0, 1] = 1/15,$$

$$\varphi_0\{p_1, p_2, p_3, p_4\}[0, 1, 0, 0] = 1/5,$$

$$\varphi_0\{p_1, p_2, p_3, p_4\}[0, 0, 1, 1] = 1/5,$$

$$\varphi_0\{p_1, p_2, p_3, p_4\}[1, 0, 0, 1] = 1/5,$$

$$\varphi_0\{p_1, p_2, p_3, p_4\}[0, 0, 0, 1] = 1/5.$$

From the above values, taking into account the definitions of the five entities as given in the previous tableau, we get:

$$\varphi_0\{P_i\}[1] = \sum_{(k_1,\ldots,k_4)\in A_i} \varphi_0\{p_1,\ldots,p_4\}[k_1,\ldots,k_4],$$

namely,

$$\varphi_0\{P_1\}[1] = 1/5, \quad \varphi_0\{P_2\}[1] = 1/5,$$

$$\varphi_0\{P_3\}[1] = 1/5, \quad \varphi_0\{P_4\}[1] = 1/5,$$

$$\varphi_0\{P_5\}[1] = 1/5,$$

which tell us that we start from the uniform distribution on the set of possible values of the decision variable P. In such a case the initial amount of uncertainty on the set of possible entities is maximum, namely,

$$H_0(P) = -\sum_{i=1}^{5} \varphi_0\{P_i\}[1] \log \varphi_0\{P_i\}[1] = \log 5 = 1.60944.$$

We are looking for the best characteristic to start with the pattern-recognition. We apply the formulas:

$$\varphi_0\{p_j/A_i\}[k] = \frac{\sum_{\{(k_1,\dots,k_4)\in A_i, k_j=k\}} \varphi_0\{p_1,\dots,p_4\}[k_1,\dots,k_4]}{\sum_{(k_1,\dots,k_4)\in A_i} \varphi_0\{p_1,\dots,p_4\}[k_1,\dots,k_4]},$$

$$\varphi_0\{p_j\}[k] = \sum_{i=1}^{5} \varphi_0\{P_i\}[1] \varphi_0\{p_j/A_i\}[k],$$

$$\varphi_1\{P_i \mid p_j\}[1 \mid k] = \frac{\varphi_0\{P_i\}[1] \varphi_0\{p_j/A_i\}[k]}{\varphi_0\{p_j\}[k]},$$

for $k \in V = \{0,1\}$, $j = 1,2,3,4$, and $i = 1,2,3,4,5$. Finally, the uncertainty on the set of available entities after checking the characteristic p_j is:

$$H_1(P \mid p_j) =$$

$$= \sum_{k\in\{1,0\}} \varphi_0\{p_j\}[k] \left(-\sum_{i=1}^{5} \varphi_1\{P_i \mid p_j\}[1 \mid k] \log \varphi_1\{P_i \mid p_j\}[1 \mid k] \right).$$

For characteristic p_1 we have:

$$\varphi_0\{p_1/A_1\}[1] = 1/3, \qquad \varphi_0\{p_1/A_2\}[1] = 0,$$
$$\varphi_0\{p_1/A_3\}[1] = 0, \qquad \varphi_0\{p_1/A_4\}[1] = 1,$$
$$\varphi_0\{p_1/A_5\}[1] = 0, \qquad \varphi_0\{p_1\}[1] = 4/15.$$

$$\varphi_0\{p_1/A_1\}[0] = 2/3, \qquad \varphi_0\{p_1/A_2\}[0] = 1,$$
$$\varphi_0\{p_1/A_3\}[0] = 1, \qquad \varphi_0\{p_1/A_4\}[0] = 0,$$
$$\varphi_0\{p_1/A_5\}[0] = 1, \qquad \varphi_0\{p_1\}[0] = 11/15.$$

$$\varphi_0\{P_1 \mid p_1\}[1 \mid 1] = 1/4, \qquad \varphi_0\{P_2 \mid p_1\}[1 \mid 1] = 0,$$
$$\varphi_0\{P_3 \mid p_1\}[1 \mid 1] = 0, \qquad \varphi_0\{P_4 \mid p_1\}[1 \mid 1] = 3/4,$$
$$\varphi_0\{P_5 \mid p_1\}[1 \mid 1] = 0.$$

$$\varphi_0\{P_1 \mid p_1\}[1 \mid 0] = 2/11, \qquad \varphi_0\{P_2 \mid p_1\}[1 \mid 0] = 3/11,$$
$$\varphi_0\{P_3 \mid p_1\}[1 \mid 0] = 3/11, \qquad \varphi_0\{P_4 \mid p_1\}[1 \mid 0] = 0,$$
$$\varphi_0\{P_5 \mid p_1\}[1 \mid 0] = 3/11.$$

The amount of uncertainty on the set of entities after checking the characteristic p_1 is:

$$H_1(P \mid p_1) = 1.1568.$$

For characteristic p_2 we have:

$$\varphi_0\{p_2/A_1\}[1] = 1, \qquad \varphi_0\{p_2/A_2\}[1] = 1,$$
$$\varphi_0\{p_2/A_3\}[1] = 0, \qquad \varphi_0\{p_2/A_4\}[1] = 0,$$
$$\varphi_0\{p_2/A_5\}[1] = 0, \qquad \varphi_0\{p_2\}[1] = 2/5.$$

$$\varphi_0\{p_2/A_1\}[0] = 0, \qquad \varphi_0\{p_2/A_2\}[0] = 0,$$
$$\varphi_0\{p_2/A_3\}[0] = 1, \qquad \varphi_0\{p_2/A_4\}[0] = 1,$$
$$\varphi_0\{p_2/A_5\}[0] = 1, \qquad \varphi_0\{p_2\}[0] = 3/5.$$

$$\varphi_0\{P_1 \mid p_2\}[1 \mid 1] = 1/2, \qquad \varphi_0\{P_2 \mid p_2\}[1 \mid 1] = 1/2,$$
$$\varphi_0\{P_3 \mid p_2\}[1 \mid 1] = 0, \qquad \varphi_0\{P_4 \mid p_2\}[1 \mid 1] = 0,$$
$$\varphi_0\{P_5 \mid p_2\}[1 \mid 1] = 0.$$

$$\varphi_0\{P_1 \mid p_2\}[1 \mid 0] = 0, \qquad \varphi_0\{P_2 \mid p_2\}[1 \mid 0] = 0,$$
$$\varphi_0\{P_3 \mid p_2\}[1 \mid 0] = 1/3, \qquad \varphi_0\{P_4 \mid p_2\}[1 \mid 0] = 1/3,$$
$$\varphi_0\{P_5 \mid p_1\}[1 \mid 0] = 1/3.$$

The amount of uncertainty on the set of entities after checking the characteristic p_2 is:

$$H_1(P \mid p_2) = 0.9364.$$

For characteristic p_3 we have:

$$\varphi_0\{p_3/A_1\}[1] = 1/3, \qquad \varphi_0\{p_3/A_2\}[1] = 0,$$
$$\varphi_0\{p_3/A_3\}[1] = 1, \qquad \varphi_0\{p_3/A_4\}[1] = 0,$$
$$\varphi_0\{p_3/A_5\}[1] = 0, \qquad \varphi_0\{p_3\}[1] = 4/15.$$

$$\varphi_0\{p_3/A_1\}[0] = 2/3, \qquad \varphi_0\{p_3/A_2\}[0] = 1,$$
$$\varphi_0\{p_3/A_3\}[0] = 0, \qquad \varphi_0\{p_3/A_4\}[0] = 1,$$
$$\varphi_0\{p_3/A_5\}[0] = 1, \qquad \varphi_0\{p_3\}[0] = 11/15.$$

$$\varphi_0\{P_1 \mid p_3\}[1 \mid 1] = 1/4, \qquad \varphi_0\{P_2 \mid p_3\}[1 \mid 1] = 0,$$
$$\varphi_0\{P_3 \mid p_3\}[1 \mid 1] = 3/4, \qquad \varphi_0\{P_4 \mid p_3\}[1 \mid 1] = 0,$$
$$\varphi_0\{P_5 \mid p_3\}[1 \mid 1] = 0.$$

$$\varphi_0\{P_1 \mid p_3\}[1 \mid 0] = 2/11, \qquad \varphi_0\{P_2 \mid p_3\}[1 \mid 0] = 3/11,$$
$$\varphi_0\{P_3 \mid p_3\}[1 \mid 0] = 0, \qquad \varphi_0\{P_4 \mid p_3\}[1 \mid 0] = 3/11,$$
$$\varphi_0\{P_5 \mid p_3\}[1 \mid 0] = 3/11.$$

The amount of uncertainty on the set of entities after checking the characteristic p_3 is:

$$H_1(P \mid p_3) = 1.1568.$$

For characteristic p_4 we have:

$$\varphi_0\{p_4/A_1\}[1] = 1, \qquad \varphi_0\{p_4/A_2\}[1] = 0,$$
$$\varphi_0\{p_4/A_3\}[1] = 1, \qquad \varphi_0\{p_4/A_4\}[1] = 1,$$
$$\varphi_0\{p_4/A_5\}[1] = 1, \qquad \varphi_0\{p_4\}[1] = 4/5.$$

$$\varphi_0\{p_4/A_1\}[0] = 0, \qquad \varphi_0\{p_4/A_2\}[0] = 1,$$
$$\varphi_0\{p_4/A_3\}[0] = 0, \qquad \varphi_0\{p_4/A_4\}[0] = 0,$$
$$\varphi_0\{p_4/A_5\}[0] = 0, \qquad \varphi_0\{p_4\}[0] = 1/5.$$

$$\varphi_0\{P_1 \mid p_4\}[1 \mid 1] = 1/4, \qquad \varphi_0\{P_2 \mid p_4\}[1 \mid 1] = 0,$$
$$\varphi_0\{P_3 \mid p_4\}[1 \mid 1] = 1/4, \qquad \varphi_0\{P_4 \mid p_4\}[1 \mid 1] = 1/4,$$
$$\varphi_0\{P_5 \mid p_4\}[1 \mid 1] = 1/4.$$

$$\varphi_0\{P_1 \mid p_4\}[1 \mid 0] = 0, \qquad \varphi_0\{P_2 \mid p_4\}[1 \mid 0] = 1,$$
$$\varphi_0\{P_3 \mid p_4\}[1 \mid 0] = 0, \qquad \varphi_0\{P_4 \mid p_4\}[1 \mid 0] = 0,$$
$$\varphi_0\{P_5 \mid p_4\}[1 \mid 0] = 0.$$

The amount of uncertainty on the set of entities after checking the characteristic p_4 is:

$$H_1(P \mid p_4) = 1.1090.$$

From the above computation we see that:

$$H_1(P \mid p_2) = \min_{j-1,\dots,4} H_1(P \mid p_j).$$

Therefore, in this problem, the pattern-recognition algorithm has to start by checking characteristic p_2. Depending on the particular value of p_2, we want now to find the next best characteristic to deal with at the second step of the algorithm.

1.1. Assume that p_2 has the value 1 with certainty, i.e., consider the circumstance:

$$Q = \text{``}\varphi_1\{p_2\}[1] = 1,\text{''}$$

The nonempty subsets $B_i = A_i/Q \subseteq A_i$ are shown in the next tableau:

B_i	k_1	k_2	k_3	k_4
B_1	1	1	0	1
	0	1	1	1
	0	1	0	1
B_2	0	1	0	0

Therefore, only the entities P_1 and P_2 remain in competition under the circumstance Q. We apply the general formulas:

$$\varphi_1\{P_i/Q\}[1] = \frac{\sum_{(k_1,\ldots,k_4)\in B_i} \varphi_0\{p_1,\ldots,p_4\}[k_1,\ldots,k_4]}{\sum_{(k_1,\ldots,k_4)\in B_1\cup\ldots\cup B_4} \varphi_0\{p_1,\ldots,p_4\}[k_1,\ldots,k_4]},$$

$$\varphi_1\{p_j/B_i\}[k] = \frac{\sum_{\{(k_1,\ldots,k_4)\in B_i, k_j=k\}} \varphi_0\{p_1,\ldots,p_4\}[k_1,\ldots,k_4]}{\sum_{(k_1,\ldots,k_4)\in B_i} \varphi_0\{p_1,\ldots,p_4\}[k_1,\ldots,k_4]},$$

$$\varphi_1\{p_j/Q\}[k] = \sum_{i=1}^{2} \varphi_1\{P_i/Q\}[1]\,\varphi_1\{p_j/B_i\}[k],$$

$$\varphi_2\{P_i/Q \mid p_j\}[1 \mid k] = \frac{\varphi_1\{P_i/Q\}[1]\,\varphi_1\{p_j/B_i\}[k]}{\varphi_1\{p_j/Q\}[k]},$$

for every $k \in V_j$, $j = 1, 3, 4$, and $i = 1, 2$. Finally, the conditional entropy of the decision variable P, under circumstance Q, taken all possible values of characteristic p_j, $(j \neq 2)$, into account is:

$$H_2(P/Q \mid p_j) =$$

$$= \sum_{k\in V_j} \varphi_1\{p_j/Q\}[k] \left(-\sum_{i=1}^{2} \varphi_2\{P_i/Q \mid p_j\}[1 \mid k] \log \varphi_2\{P_i/Q \mid p_j\}[1 \mid k] \right).$$

We obtain:

$$\varphi_1\{P_1/Q\}[1] = 1/2, \quad \varphi_1\{P_2/Q\}[1] = 1/2.$$

For the characteristic p_1 we have:

$$\varphi_0\{p_1/B_1\}[1] = 1/3, \quad \varphi_0\{p_1/B_2\}[1] = 0,$$

$$\varphi_0\{p_1/B_1\}[0] = 2/3, \quad \varphi_0\{p_1/B_2\}[0] = 1,$$

$$\varphi_1\{p_1/Q\}[1] = 1/6, \quad \varphi_1\{p_1/Q\}[0] = 5/6,$$

$$\varphi_1\{P_1/Q \mid p_1\}[1 \mid 1] = 1, \qquad \varphi_1\{P_2/Q \mid p_1\}[1 \mid 1] = 0,$$
$$\varphi_1\{P_1/Q \mid p_1\}[1 \mid 0] = 2/5, \qquad \varphi_1\{P_2/Q \mid p_1\}[1 \mid 0] = 3/5,$$

$$H_2(P/Q \mid p_1) = \frac{1}{6}(-1 \log 1 - 0 \log 0) + \frac{5}{6}(-\frac{2}{5} \log \frac{2}{5} - \frac{3}{5} \log \frac{3}{5}) = 0.56084,$$

where $-0 \log 0 = 0$, obtained extending the function $-x \log x$ by continuity at the origin.

For the characteristic p_3 we have:

$$\varphi_0\{p_3/B_1\}[1] = 1/3, \quad \varphi_0\{p_3/B_2\}[1] = 0,$$

$$\varphi_0\{p_3/B_1\}[0] = 2/3, \quad \varphi_0\{p_3/B_2\}[0] = 1,$$

$$\varphi_1\{p_3/Q\}[1] = 1/6, \quad \varphi_1\{p_3/Q\}[0] = 5/6,$$

$$\varphi_1\{P_1/Q \mid p_3\}[1 \mid 1] = 1, \qquad \varphi_1\{P_2/Q \mid p_3\}[1 \mid 1] = 0,$$

$$\varphi_1\{P_1/Q \mid p_3\}[1 \mid 0] = 2/5, \qquad \varphi_1\{P_2/Q \mid p_3\}[1 \mid 0] = 3/5,$$

$$H_2(P/Q \mid p_3) = 0.56084,$$

For the characteristic p_4 we have:

$$\varphi_0\{p_4/B_1\}[1] = 1, \quad \varphi_0\{p_4/B_2\}[1] = 0,$$

$$\varphi_0\{p_4/B_1\}[0] = 0, \quad \varphi_0\{p_4/B_2\}[0] = 1,$$

$$\varphi_1\{p_4/Q\}[1] = 1/2, \quad \varphi_1\{p_1/Q\}[0] = 1/2,$$

$$\varphi_1\{P_1/Q \mid p_4\}[1 \mid 1] = 1, . \qquad \varphi_1\{P_2/Q \mid p_4\}[1 \mid 1] = 0,$$

$$\varphi_1\{P_1/Q \mid p_4\}[1 \mid 0] = 0, \qquad \varphi_1\{P_2/Q \mid p_4\}[1 \mid 0] = 1,$$

$$H_2(P/Q \mid p_4) = 0.$$

From the above computation we see that:

$$H_2(P/Q \mid p_4) = \min_{j=1,3,4} H_2(P/Q \mid p_j).$$

Therefore, if the characteristic p_2 has the value 1, then the next characteristic to be checked is p_4. Let us examine what happens if this last characteristic has either the value 1 or the value 0.

2.1. Let us continue our analysis under the circumstance:

$$Q : \text{``}\varphi_1\{p_2\}[1] = 1, \varphi_2\{p_4\}[1] = 1.\text{''}$$

Under this new circumstance, there is only one corresponding subset, namely $C_1 = A_1/Q$, shown below:

C_i	k_1	k_2	k_3	k_4
C_1	1	1	0	1
	0	1	1	1
	0	1	0	1

in which case we have $\varphi_2\{P_1/Q\}[1] = 1$ which shows that, under the circumstance Q, only entity P_1 is possible.

2.2. Let us continue our analysis under the circumstance:

$$Q : \text{``}\varphi_1\{p_2\}[1] = 1, \varphi_2\{p_4\}[0] = 1.\text{''}$$

Under this new circumstance, there is only one corresponding subset, namely $C_2 = A_2/Q$, shown below,

C_i	k_1	k_2	k_3	k_4
C_2	0	1	0	0

in which case we have $\varphi_2\{P_2/Q\}[1] = 1$ which shows that, under the circumstance Q, only entity P_2 is possible.

1.2. Going back to the first step of the algorithm, assume that p_2 has the value 0 with certainty, i.e., consider the circumstance:

$$Q = \text{``}\varphi_1\{p_2\}[0] = 1,\text{''}$$

The nonempty subsets $B_i = A_i/Q \subseteq A_i$ are shown in the next tableau:

B_i	k_1	k_2	k_3	k_4
B_3	0	0	1	1
B_4	1	0	0	1
B_5	0	0	0	1

Therefore, only the entities P_3, P_4, and P_5 remain in competition under the new circumstance Q. Applying again the formulas mentioned at step 1.1 discussed above, we get:

$$\varphi_1\{P_3/Q\}[1] = 1/3, \varphi_1\{P_4/Q\}[1] = 1/3, \varphi_1\{P_5/Q\}[1] = 1/3,$$

$$\varphi_1\{p_1/B_3\}[1] = 0, \ \varphi_1\{p_1/B_4\}[1] = 1, \ \varphi_1\{p_1/B_5\}[1] = 0,$$

$$\varphi_1\{p_1/B_3\}[0] = 1, \ \varphi_1\{p_1/B_4\}[0] = 0, \ \varphi_1\{p_1/B_5\}[0] = 1,$$

$$\varphi_1\{p_1/Q\}[1] = 1/3, \quad \varphi_1\{p_1/Q\}[0] = 2/3,$$

$$\varphi_2\{P_3/Q \mid p_1\}[1 \mid 1] = 0, \qquad \varphi_2\{P_3/Q \mid p_1\}[1 \mid 0] = 1/2,$$
$$\varphi_2\{P_4/Q \mid p_1\}[1 \mid 1] = 1, \qquad \varphi_2\{P_4/Q \mid p_1\}[1 \mid 0] = 0,$$
$$\varphi_2\{P_5/Q \mid p_1\}[1 \mid 1] = 0, \qquad \varphi_2\{P_5/Q \mid p_1\}[1 \mid 0] = 1/2,$$

$$H_2(P/Q \mid p_1) = \frac{1}{3}(-0 \log 0 - 1 \log 1 - 0 \log 0) +$$

$$+ \frac{2}{3}(-\frac{1}{2} \log \frac{1}{2} - 0 \log 0 - \frac{1}{2} \log \frac{1}{2}) = 0.4621.$$

$$\varphi_1\{p_3/B_3\}[1] = 1, \ \varphi_1\{p_3/B_4\}[1] = 0, \ \varphi_1\{p_3/B_5\}[1] = 0,$$

$$\varphi_1\{p_3/B_3\}[0] = 0, \ \varphi_1\{p_3/B_4\}[0] = 1, \ \varphi_1\{p_3/B_5\}[0] = 1,$$

$$\varphi_1\{p_3/Q\}[1] = 1/3, \quad \varphi_1\{p_3/Q\}[0] = 2/3,$$

$$\varphi_2\{P_3/Q \mid p_3\}[1 \mid 1] = 1, \qquad \varphi_2\{P_3/Q \mid p_3\}[1 \mid 0] = 0,$$
$$\varphi_2\{P_4/Q \mid p_3\}[1 \mid 1] = 0, \qquad \varphi_2\{P_4/Q \mid p_3\}[1 \mid 0] = 1/2,$$
$$\varphi_2\{P_5/Q \mid p_3\}[1 \mid 1] = 0, \qquad \varphi_2\{P_5/Q \mid p_3\}[1 \mid 0] = 1/2,$$

$$H_2(P/Q \mid p_3) = \frac{1}{3}(-1 \log 1 - 0 \log 0 - 0 \log 0) +$$

$$+ \frac{2}{3}(-0 \log 0 - \frac{1}{2} \log \frac{1}{2} - \frac{1}{2} \log \frac{1}{2}) = 0.4621.$$

$$\varphi_1\{p_4/B_3\}[1] = 1, \ \varphi_1\{p_4/B_4\}[1] = 1, \ \varphi_1\{p_4/B_5\}[1] = 1,$$

$$\varphi_1\{p_4/B_3\}[0] = 0, \ \varphi_1\{p_4/B_4\}[0] = 0, \ \varphi_1\{p_4/B_5\}[0] = 0,$$

$$\varphi_1\{p_4/Q\}[1] = 1, \quad \varphi_1\{p_4/Q\}[0] = 0,$$

$$\varphi_2\{P_3/Q \mid p_4\}[1 \mid 1] = 1/3,$$
$$\varphi_2\{P_4/Q \mid p_4\}[1 \mid 1] = 1/3,$$
$$\varphi_2\{P_5/Q \mid p_4\}[1 \mid 1] = 1/3,$$

$$H_2(P/Q \mid p_4) = 1[3 \times (-\frac{1}{3} \log \frac{1}{3})] = 1.0986.$$

Therefore, if the characteristic p_2 has the value 0, then the best characteristic to check at the second step of the pattern-recognition algorithm is either p_1 or p_3, because:

$$H_2(P/Q \mid p_1) = H_2(P/Q \mid p_3) = \min_{1 \leq j \leq 4, j \neq 2} H_2(P/Q \mid p_j).$$

The tie between p_1 and p_3 induces two equivalent pattern-recognition algorithms corresponding to the two different choices we can make at this step. Let us choose the characteristic p_1 and continue the elaboration of the pattern-recognition algorithm. If we choose p_3 instead, a similar approach can be implemented.

2.3. Let us take the circumstance:

$$Q: \text{``}\varphi_1\{p_2\}[0] = 1, \quad \varphi_2\{p_1\}[1] = 1.\text{''}$$

Taking $C_i = A_i/Q$, the only nonempty such subset is:

C_i	k_1	k_2	k_3	k_4
C_4	1	0	0	1

and, consequently,

$$\varphi_2\{P_4/Q\}[1] = 1.$$

2.4. Let us take the circumstance:

$$Q: \text{``}\varphi_1\{p_2\}[0] = 1, \quad \varphi_2\{p_1\}[0] = 1.\text{''}$$

Taking $C_i = A_i/Q$, the only nonempty such subsets are:

C_i	k_1	k_2	k_3	k_4
C_3	0	0	1	1
C_5	0	0	0	1

We have:

$$\varphi_2\{P_3/Q\}[1] = 1/2, \qquad \varphi_2\{P_5/Q\}[1] = 1/2,$$
$$\varphi_2\{p_3/C_3\}[1] = 1, \qquad \varphi_2\{p_3/C_5\}[1] = 0,$$
$$\varphi_2\{p_3/C_3\}[0] = 0, \qquad \varphi_2\{p_3/C_5\}[0] = 1,$$

$$\varphi_2\{p_3/Q\}[1] = 1/2, \qquad \varphi_2\{p_3/Q\}[0] = 1/2,$$
$$\varphi_3\{P_3/Q \mid p_3\}[1 \mid 1] = 1, \qquad \varphi_3\{P_3/Q \mid p_3\}[1 \mid 0] = 0,$$
$$\varphi_3\{P_5/Q \mid p_3\}[1 \mid 1] = 0, \qquad \varphi_3\{P_5/Q \mid p_3\}[1 \mid 0] = 1,$$

$$H_3(P/Q \mid p_3) = \frac{1}{2}(-1 \log 1 - 0 \log 0) + \frac{1}{2}(-0 \log 0 - 1 \log 1) = 0.$$

$$\varphi_2\{p_4/C_3\}[1] = 1, \qquad \varphi_2\{p_4/C_5\}[1] = 1,$$
$$\varphi_2\{p_4/C_3\}[0] = 0, \qquad \varphi_2\{p_4/C_5\}[0] = 0,$$

$$\varphi_2\{p_4/Q\}[1] = 1, \quad \varphi_2\{p_4/Q\}[0] = 0,$$

$$\varphi_3\{P_3/Q \mid p_4\}[1 \mid 1] = 1/2,$$
$$\varphi_3\{P_5/Q \mid p_4\}[1 \mid 1] = 1/2,$$

$$H_3(P/Q \mid p_4) = 1[2 \times (-\frac{1}{2} \log \frac{1}{2})] = 0.69315.$$

Therefore, if p_2 has the value 0 at the first step and p_1 has the value 0 at the second step, then the next characteristic to be checked at the third step of the algorithm is p_3, because:

$$H_3(P/Q \mid p_3) = \min_{j=3,4} H_3(P/Q \mid p_j).$$

3.1. Under the new circumstance:

$$Q : \text{``} \varphi_1\{p_2\}[0] = 1, \quad \varphi_2\{p_1\}[0] = 1, \quad \varphi_3\{p_3\}[1] = 1,\text{''}$$

taking the subsets $D_i = A_i/Q$, the only such subset that is not empty is shown in the following tableau:

D_i	k_1	k_2	k_3	k_4
D_3	0	0	1	1

corresponding to the only entity:

$$\varphi_3\{P_3/Q\}[1] = 1.$$

3.2. Under the new circumstance:

$$Q : \text{``} \varphi_1\{p_2\}[0] = 1, \quad \varphi_2\{p_1\}[0] = 1, \quad \varphi_3\{p_3\}[0] = 1,\text{''}$$

taking the subsets $E_i = A_i/Q$, the only such subset that is not empty is shown in the following tableau:

E_i	k_1	k_2	k_3	k_4
E_5	0	0	0	1

corresponding to the only entity:

$$\varphi_3\{P_5/Q\}[1] = 1.$$

Summarizing the results of our analysis, we obtain the following pattern-recognition algorithm:

$$p_2? \begin{cases} 1 \longrightarrow p_4? \begin{cases} 1 \Longrightarrow P_1 \\ \\ 0 \Longrightarrow P_2 \end{cases} \\ \\ 0 \longrightarrow p_1? \begin{cases} 1 \Longrightarrow P_4 \\ \\ 0 \longrightarrow p_3? \begin{cases} 1 \Longrightarrow P_3 \\ \\ 0 \Longrightarrow P_5 \end{cases} \end{cases} \end{cases}$$

If we choose to check p_3 instead of p_1 at the second step of the algorithm on the path corresponding to the value 0 of p_2, we obtain the equivalent solution:

$$p_2? \begin{cases} 1 \longrightarrow p_4? \begin{cases} 1 \Longrightarrow P_1 \\ \\ 0 \Longrightarrow P_2 \end{cases} \\ \\ 0 \longrightarrow p_3? \begin{cases} 1 \Longrightarrow P_3 \\ \\ 0 \longrightarrow p_1? \begin{cases} 1 \Longrightarrow P_4 \\ \\ 0 \Longrightarrow P_5 \end{cases} \end{cases} \end{cases}$$

Let us emphasize that the pattern-recognition algorithm contains all possible paths but, when it is applied to a particular case, we follow the indications of only one of these possible paths. We can see that never more than three characteristics have to be determined and there are cases when the values of only two of them are really needed. If we calculate the mean number of characteristic checks, we obtain:

$$2 \times \frac{1}{5} + 2 \times \frac{1}{5} + 3 \times \frac{1}{5} + 3 \times \frac{1}{5} = 2.4,$$

instead of 4, which happens when all character istics are checked before identifying the corresponding entity. The simplification is more obvious, even striking, when the problem involves more entities and characteristics. The pattern-recognition algorithm based on the relative logic is a useful tool in coping with complexity, one of the main difficulties in natural and artificial intelligence.

5.2. Weighting credibilities

As the relative logic deals with probability or credibility distributions on the set of possible values of a sentence, it is necessary to discuss briefly something about ways of calculating such distributions both from static (i.e., when time is frozen) and evolutionary (i.e., when time flows) points of view.

1. The static approach. At a given time instant, sometimes the probability distribution is obtained using objective relative frequencies based on past observations. When there are no such relative frequencies available, subjective credibility distributions may be used. Thus, if we have no information about the possible outcomes and we have no reason to discriminate between different possible outcomes, we apply Laplace's Principle of Insufficient Reason taking the outcomes to be equally likely, i.e., assuming the uniform distribution on the possible outcomes. If the only information available consists of the mean values of some random variables depending on an unknown probability distribution, we may apply the Maximum Entropy Principle, according to which we construct the unique probability distribution that maximizes Shannon's entropy, a common measure of the amount of uncertainty, subject to the given mean values. Such a probability distribution is the most unbiased distribution subject to the given mean values. Thus, if $V = \{h_1, \ldots, h_m\}$ is the set of possible logical values of a sentence p

and the only information available is the mean value:

$$\mu = \sum_{i=1}^{m} x_i \varphi\{p\}[h_i],$$

then, the probability distribution:

$$\varphi\{p\}[h_i] > 0, \quad \sum_{i=1}^{m} \varphi\{p\}[h_i] = 1$$

that maximizes the entropy:

$$H(\varphi\{p\}) = -\sum_{i=1}^{m} \varphi\{p\}[h_i] \log \varphi\{p\}[h_i],$$

i.e., the probability distribution containing the maximum amount of uncertainty, or equivalently, the most unbiased probability distribution, subject to the given mean value μ, is:

$$\varphi\{p\}[h_i] = \frac{1}{\Phi(\alpha)} \exp(-\alpha\, x_i), \quad (i = 1, \ldots, m),$$

where:

$$\Phi(\alpha) = \sum_{i=1}^{m} \exp(-\alpha\, x_i),$$

and α is the unique solution of the equation:

$$\frac{d \log \Phi(\alpha)}{d\alpha} = -\mu.$$

2. The evolutionary approach. Two rules for calculating successive probability/credibility distributions are common in applications. One of them, the Markov chain rule, starts from a prior probability distribution on a set of possible values and calculates the posterior probability distributions of these values at successive time instants in a linear way, using stochastic matrices representing transition probabilities from an arbitrary value to an arbitrary value. Often, a Markov chain describes the probabilistic internal evolution of an isolated, closed system and the equilibrium is reached by increasing the global amount of uncertainty on the set of possible states of the respective system. The behaviour of the molecules of a gas in a closed room is an example of such a probabilistic model. A second rule for calculating successive probabilities, the Bayes' rule, on the other hand, is applied for evaluating the posterior credibilities of a set of available logical values or hypotheses, weighting the prior credibility distribution, in a nonlinear way, by using stochastic matrices representing the prediction probabilities/credibilities of the outcomes of an auxiliary experiment conditioned by the available hypotheses. While the Markov chains deal mainly with objective probabilities and focus only on a set of internal states of the respective system, the Bayesian models often involve subjective probabilities, or credibilities, and open up the system of available hypotheses, correlating them with outcomes of an external auxiliary experiment. A drawback of Bayesian approach, however, is that once an available hypothesis has been rejected, a new evidence cannot reinstate it, contrary to what happens in many real cases. The objective of this section is to propose a more general way of weighting the prior probability/credibility distribution and subsequent posterior probabilities/credibilities of the available logical values or hypotheses in order to overcome this difficulty. Both the Markov chain rule and the Bayes' rule are obtained as particular cases. Special cases are obtained when the weighting process is done in a multiplicative way, whose stationary variant includes the reinforcement process so frequently used in automatic learning, or when the available logical values or hypotheses are temporally incompatible. An example will deal with the process of weighting the credibilities of available hypotheses in a detective story, based on the accumulation of successive pieces of evidence, where the corresponding weights reflect the motive and opportunity of the suspects involved in a murder case. The credibility distribution obtained from the corresponding weighting process allows us to calculate the values of the belief, plausibility, and ambiguity functions for different sets of suspects during the investigation process. In general, both in the two-valued and in the multi-valued logics, the set of possible logical values is kept fixed. There are, however, many real life situations when this is not the case and the set of possible logical values changes in time, essentially depending on the accumulation of new evidence which suppliments or contradicts previous evidence available. Thus, if the main sentence we are dealing with is "Who is the murderer?" and the set of possible logical values is the set of suspects, then this set could change in time, some of its characters being cleared up, therefore eliminated from the set, whereas other new characters join the set, depending on the new evidence or circumstances becoming available.

3. The sequential process of weighting credibilities. Let p be a sentence and V_0 a finite set of logical values or available hypotheses at time instant 0 and:

$$\{\varphi_0\{p\}[h]; h \in V_0\}$$

a prior probability (or credibility) distribution on V_0. For each logical value or available hypothesis $h \in V_0$, let:

$$\{u_{0,1}(\tilde{h} \mid h); \tilde{h} \in V_1\}$$

be a set of subjective or objective nonnegative weights assigned to the logical values or available hypotheses of V_1 at time instant 1, given that the logical value or hypothesis h, available at the initial time instant 0, is confronted with the new evidence available at time instant 1. The set V_1 may coincide to V_0, have only a nonempty intersection with V_0, or be completely different from V_0. The weights are induced by some new evidence provided at time instant 1 which determines the decision maker to update, or rethink, the credibility of the logical values or available hypotheses. The credibilities of the logical values or hypotheses from V_1, at time instant 1, are:

$$\varphi_1\{p\}[\tilde{h}] = \frac{\sum_{h \in V_0} u_{0,1}(\tilde{h} \mid h)\, \varphi_0\{p\}[h]}{\sum_{\tilde{h} \in V_1} \sum_{h \in V_0} u_{0,1}(\tilde{h} \mid h)\, \varphi_0\{p\}[h]}.$$

In general, let V_n be a finite set of logical values or hypotheses available at time instant n and, for each logical value or hypothesis $h \in V_0$, and let:

$$\{u_{0,n}(\tilde{h} \mid h), \tilde{h} \in V_n\}$$

be a set of subjective or objective nonnegative weights of the logical values or hypotheses of V_n, at time instant n, assuming that the logical value or hypothesis h, available at time instant 0, is confronted with the evidence provided during the time interval $[0, n]$. The credibilities of the logical values or hypotheses from V_n, at time instant n, are:

$$\varphi_n\{p\}[\tilde{h}] = \frac{\sum_{h \in V_0} u_{0,n}(\tilde{h} \mid h)\, \varphi_0\{p\}[h]}{\sum_{\tilde{h} \in V_n} \sum_{h \in V_0} u_{0,n}(\tilde{h} \mid h)\, \varphi_0\{p\}[h]}.$$

The amount of uncertainty on the set of logical values or available hypotheses V_n at time n is measured by Shannon's entropy:

$$H(V_n) = - \sum_{\tilde{h} \in V_n} \varphi_n\{p\}[\tilde{h}] \log \varphi_n\{p\}[\tilde{h}],$$

where log is the natural logarithm.

4. Special cases: 1.) The weighting process is multiplicative when:

$$u_{0,n}(\tilde{h} \mid h) = \sum_{\bar{h} \in V_m} u_{m,n}(\tilde{h} \mid \bar{h}) \, u_{0,m}(\bar{h} \mid h),$$

for each $h \in V_0$, $\tilde{h} \in V_n$, and $0 < m < n$. In such a case, the weights:

$$u_{k,k+1}(\tilde{h} \mid h), \quad (h \in V_k, \tilde{h} \in V_{k+1}), \quad (k = 0, 1, \ldots),$$

completely determine the weights $u_{0,n}(\tilde{h} \mid h)$. A multiplicative weighting process is a reinforcement process, or a stationary multiplicative weighting process, if:

$$u_{k,k+1}(\tilde{h} \mid h) = u(\tilde{h} \mid h), \quad (k = 0, 1, \ldots).$$

2.) A weighting process is independent on the initial credibilities, or a prejudice-free weighting process, if:

$$u_{0,n}(\tilde{h} \mid h) = u_{0,n}(\tilde{h}),$$

for every $h \in V_0$, $\tilde{h} \in V_n$, and n, in which case the new probabilities/credibilities:

$$\varphi_n\{p\}[\tilde{h}] = \frac{u_{0,n}(\tilde{h})}{\sum_{\tilde{h} \in V_n} u_{0,n}(\tilde{h})}, \quad (\tilde{h} \in V_n),$$

depend only on the evidence available during the time interval $[0, n]$.

5. Markov chains. Assume that:

$$V_0 = V_1 = \ldots = V_{n-1} = V_n = \ldots = V$$

are the possible states of a system, the same at each time instant, and let us consider a multiplicative weighting process where:

$$u_{n-1,n}(\tilde{h} \mid h)$$

is the transition probability from state h at time instant $n-1$ to state $\tilde{h}$ at time instant n. Then:

$$u_{n-1,n}(\tilde{h} \mid h) \geq 0, \quad \sum_{\tilde{h} \in V} u_{n-1,n}(\tilde{h} \mid h) = 1,$$

for every $h \in V$, and we have a Markov chain on V, according to:

$$\varphi_n\{p\}[\tilde{h}] = \sum_{h \in V} u_{n-1,n}(\tilde{h} \mid h)\, \varphi_{n-1}\{p\}[h],$$

for each $\tilde{h} \in V$. This Markov chain is stationary if:

$$u_{n-1,n}(\tilde{h} \mid h) = u(\tilde{h} \mid h),$$

where the matrix:

$$u = \{u(\tilde{h} \mid h); \tilde{h} \in V, h \in V\}$$

is a stochastic matrix, in which case:

$$\varphi_n\{p\}[\tilde{h}] = \sum_{h \in V} u^{(n)}(\tilde{h} \mid h)\, \varphi_0\{p\}[h], \quad (\tilde{h} \in V),$$

where $u^{(n)}$ is the n-th power of matrix u. It is well known that if some power of u has strictly positive entries, then there is a unique probability distribution:

$$\{\varphi\{p\}[h]; h \in V\}, \quad \varphi\{p\}[\tilde{h}] = \sum_{h \in V} u(\tilde{h} \mid h)\, \varphi\{p\}[h],$$

such that, for any prior probability distribution $\{\varphi_0\{p\}[h]; h \in V\}$, we have:

$$\varphi_\infty\{p\}[\tilde{h}] = \lim_{n \to +\infty} \varphi_n\{p\}[\tilde{h}] = \varphi\{p\}[\tilde{h}], \quad (\tilde{h} \in V).$$

Generally, a Markov chain describes the evolution of closed systems with random internal dynamics which, under special circumstances, could tend towards a steady-state condition. Such an internal evolution could induce an increase of the mean amount of uncertainty on the set V of possible states of the system.

6. The case of temporally incompatible hypotheses. Assume again that:

$$V_0 = V_1 = \ldots = V_{n-1} = V_n = \ldots = V.$$

The weighting process deals with temporally incompatible logical values or hypotheses when the weights, at any time instant n, are such that we have $u_{0,n}(\tilde{h} \mid h) = 0$ if $\tilde{h} \neq h$. In such a case, the weighting process of the credibilities of the logical values or available hypotheses is based only on weights of the form:

$$\{u_{0,n}(h \mid h); h \in V\},$$

which means that examining a logical value or available hypothesis $h \in V$ with respect to the new evidence at time instant n may change only the credibility of the logical value or hypothesis h itself. Let us consider a multiplicative weighting process with temporally incompatible logical values or hypotheses and denote:

$$u_{n-1,n}(h \mid h) \quad \text{by} \quad u_{n-1,n}(h).$$

We get the probability distribution at time 1:

$$\varphi_1\{p\}[h] = \frac{u_{0,1}(h)\varphi_0\{p\}[h]}{\sum_{h \in V} u_{0,1}(h)\,\varphi_0\{p\}[h]}, \quad (h \in V).$$

By iterating, the credibilities of the logical values or hypotheses from V at time instant n become:

$$\varphi_n\{p\}[h] = \frac{u_{n-1,n}(h)\,\varphi_{n-1}\{p\}[h]}{\sum_{h \in V} u_{n-1,n}(h)\,\varphi_{n-1}\{p\}[h]} =$$

$$= \frac{\prod_{k=1}^n u_{k-1,k}(h)\,\varphi_0\{p\}[h]}{\sum_{h \in V} \prod_{k=1}^n u_{k-1,k}(h)\,\varphi_0\{p\}[h]}, \quad (h \in V).$$

In the stationary case, if:

$$u_{k-1,k}(h) = u(h), \quad (k = 1, 2, \ldots),$$

the set of logical values or available hypotheses V may be decomposed into three disjoint subsets, namely: $V = \hat{V} \cup \bar{V} \cup \tilde{V}$, where:

$$\hat{V} = \{h; u(h) = 0\},$$

$$\bar{V} = \{h; u(h) \neq 0, u(h) \neq \max_h u(h)\},$$

$$\tilde{V} = \{h; u(h) = \max_h u(h)\}.$$

In this classification, $\hat{V}$ is the set of logical values or hypotheses that are rejected from the beginning, $\bar{V}$ is the set of logical values or hypotheses that are rejected asymptotically, and $\tilde{V}$ is the set of logical values or hypotheses that are confirmed by the weighting process. Indeed, introducing:

$$u_{k-1,k}(h) = u(h), \quad (k = 1, 2, \ldots),$$

we get the credibility of the logical values or available hypotheses at time instant n to be:

$$\varphi_n\{p\}[h] = \frac{(u(h))^n \, \varphi_0\{p\}[h]}{\sum_{h \in V}(u(h))^n \, \varphi_0\{p\}[h]}.$$

If $h \in \hat{V}$, then obviously,

$$\varphi_n\{p\}[h] = 0,$$

for all values of n.

Now, for any $h \in \bar{V} \cup \tilde{V}$, we have:

$$\varphi_n\{p\}[h] = \varphi_0\{p\}[h] \left/ \left[\sum_{\bar{h} \in \bar{V}} \left(\frac{u(\bar{h})}{u(h)} \right)^n \varphi_0\{p\}[\bar{h}] + \sum_{\tilde{h} \in \tilde{V}} \left(\frac{u(\tilde{h})}{u(h)} \right)^n \varphi_0\{p\}[\tilde{h}] \right] \right. .$$

If $h \in \bar{V}$, then there exists:

$$\varphi_\infty\{p\}[h] = \lim_{n \to +\infty} \varphi_n\{p\}[h] = 0,$$

because for any $\tilde{h} \in \tilde{V}$, we have:

$$\frac{u(\tilde{h})}{u(h)} > 1.$$

If $h \in \tilde{V}$, then there exists:

$$\varphi_\infty\{p\}[h] = \lim_{n \to +\infty} \varphi_n\{p\}[h] = \frac{\varphi_0\{p\}[h]}{\sum_{\tilde{h} \in \tilde{V}} \varphi_0\{p\}[\tilde{h}]},$$

because, for any $\bar{h} \in \bar{V}$, we have:

$$\frac{u(\bar{h})}{u(h)} < 1.$$

Remarks: 1^0. In general, if:

$$\prod_{k=1}^{+\infty} u_{k-1,k}(h) = \lim_{n \to +\infty} \prod_{k=1}^{n} u_{k-1,k}(h) = \lambda(h),$$

then, we get:

$$\varphi_\infty\{p\}[h] = \lim_{n \to +\infty} \varphi_n\{p\}[h] = \frac{\lambda(h)\,\varphi_0\{p\}[h]}{\sum_{h \in V} \lambda(h)\,\varphi_0\{p\}[h]}.$$

As the weights $\{u_{k-1,k}(h); k = 1, 2, \ldots\}$ are positive numbers, the product:

$$\prod_{k=1}^{\infty} u_{k-1,k}(h)$$

converges if and only if the series:

$$\sum_{k=1}^{\infty} [u_{k-1,k}(h) - 1]$$

converges.

2^0. If we have:

$$u_{k-1,k}(h) = u(h), \quad (k = 1, 2, \ldots),$$

then this process of weighting the credibilities is characterized by the decrease of the mean amount of uncertainty on the set of logical values or available hypotheses.

7. A generalization of the Bayes' rule. Assume that the experimental evidence for testing our logical values or available hypotheses at different time instants is represented by a sequence of outcomes of an auxiliary experiment, namely,

$$\{d_1, d_2, \ldots, d_{n-1}, d_n, \ldots\}$$

and let B_n be the entire body of evidence available at time instant n, namely,

$$B_n = \{d_1, d_2, \ldots, d_n\}.$$

Let us consider a weighting process whose weights

$$u_{0,n}(\tilde{h}, B_n \mid h)$$

depend on the body of evidence available during the time interval $[0, n]$, in which case the probability distribution on the set of logical values or available hypotheses at time n is:

$$\varphi_n\{p\}[\tilde{h}] = \frac{\sum_{h \in V_0} u_{0,n}(\tilde{h}, B_n \mid h)\, \varphi_0\{p\}[h]}{\sum_{\tilde{h} \in V_n} \sum_{h \in V_0} u_{0,n}(\tilde{h}, B_n \mid h)\, \varphi_0\{p\}[h]}.$$

This formula is a generalization of the Bayes' formalism from decision theory. If, in particular, such a weighting process has the same set of available hypotheses at any time, namely, $V_n = V$, $(n = 0, 1, \ldots)$, and is with temporally incompatible logical values or hypotheses, i.e.,

$$u_{0,n}(\tilde{h}, B_n \mid h) = \begin{cases} 0, & \text{if } \tilde{h} \neq h, \\ \varphi_n\{B_n \mid h\}[1 \mid 1], & \text{if } \tilde{h} = h, \end{cases}$$

where $\varphi_n\{B_n \mid h\}[1 \mid 1]$ is the prediction probability of the body of evidence B_n, given the logical value or hypothesis h at the initial moment 0. In such a case, the above formula for getting the probability distribution of the logical values or hypotheses at time n becomes the well known classic Bayes' rule for calculating the posterior probability distribution of the available hypotheses from the prior probability distribution and the prediction probabilities, namely,

$$\varphi_n\{p\}[h] = \frac{\varphi_n\{B_n \mid h\}[1 \mid 1]\, \varphi_0\{p\}[h]}{\sum_{h \in V} \varphi_n\{B_n \mid h\}[1 \mid 1]\, \varphi_0\{p\}[h]}.$$

The generalization of the Bayes' formalism discussed here, allows us to reinstate later a logical value or hypothesis previously rejected, due to some new evidence at hand, and/or to consider new logical values or hypotheses induced by the body of evidence provided by new pieces of information, new facts, or new outcomes of an auxiliary experiment.

Let us notice that in the case of a prejudice-free weighting process induced by evidence, we have:

$$u_{0,n}(\tilde{h}, B_n \mid h) = u_{0,n}(\tilde{h}, B_n),$$

for every $h \in V_0$, $\tilde{h} \in V_n$, and n. We get:

$$\varphi_n\{p\}[\tilde{h}] = \frac{u_{0,n}(\tilde{h}, B_n)}{\sum_{\tilde{h} \in V_n} u_{0,n}(\tilde{h}, B_n)}.$$

The initial probability/credibility distribution:

$$\{\varphi_0\{p\}[h]; h \in V_0\},$$

is either given, or calculated by applying the formula:

$$\varphi_0\{p\}[h] = \frac{u_0(h, d_0)}{\sum_{h \in V_0} u_0(h, d_0)},$$

where $u_0(h, d_0)$ is the initial weight of the logical value or hypothesis $h \in V_0$, compatible with the initial piece of evidence d_0.

8. Example. Let us refer again to Agatha Christie's detective story *Lord Edgware dies*, analysed in section 4.3, and look at it from another angle. The main characters are: w = Jane Wilkinson, an actress, the wife of Lord Edgware; c = Carlotta Adams, an actress who can perfectly impersonate w; m = Bryan Martin, an actor who loves w and is a friend of c; r = Captain Ronald Marsh, the nephew of Lord Edgware, in great need of money; s = Miss Carroll, Lord Edgware's secretary; g = Geraldine Marsh, Lord Edgware's daughter, who hated her father and is very fond of r; a = the aristocrat Duke of Merton whom w wants to marry; b = Lord Edgware's handsome butler. The case is investigated by the detective Hercule Poirot.

Let V_i be the set of suspects to be potential murderers of Lord Edgware at the i-th step of the investigation, where a character is suspected to be a potential murderer if, according to the evidence gathered until that moment, he/she had motive or/and opportunity for killing Lord Edgware. In his analysis, at each step of his investigation, Hercule Poirot applies Laplace's cautious Principle of Insufficient Reason, according to which, if there is no

reason to discriminate between the possible outcomes, then the best strategy
is to take them as being equally likely. Let us follow Hercule Poirot's steps in
investigating the case. The available hypotheses are individual suspects, or
groups of suspects, at different steps of the investigation. Thus, if we denote
by $X = \{w, c, m, r, s, g, a, b\}$ the set of characters, then, at any time n, the
set of suspects is:

$$V_n \subseteq \mathcal{P}(X),$$

where $\mathcal{P}(X)$ is the class of all subsets of the set X. The weighting process
involved is prejudice-free and, consequently, it is based on the formula for
calculating the credibility $\varphi_n\{p\}[\tilde{h}]$, where the weight $u_{0,n}(\tilde{h}, B_n)$ is the sum
of the degree of motive and the degree of opportunity associated to the
suspect, or group of suspects, denoted by $\tilde{h}$, taking the body of evidence
B_n into account. For an individual suspect, the degree of motive is 1 if the
respective suspect had a motive for killing Lord Edgware and 0 if he/her
had not. For a group of suspects, the degree of motive is the average of
the degrees of motive of the suspects making up the group. The same thing
for the degree of opportunity. Finally, using $\varphi_n\{p\}$ as a basic probability
assignment, we calculate, at each step, the amount of uncertainty, measured
by Shannon's entropy, together with the belief, plausibility, and ambiguity,
associated to each suspect or group of suspects, according to the formulas:

$$Bel_n(A) = \sum_{C \subseteq A} \varphi_n\{p\}[C],$$

$$Pl_n(A) = \sum_{C \cap A \neq \emptyset} \varphi_n\{p\}[C],$$

$$Amb_n(A) = Pl_n(A) - Bel_n(A),$$

where A and C are subsets of X. The investigation process consists of the
following steps:

Step 0: d_0: "Lord Edgware is murdered." Table 1 summarizes the list
of initial suspects given the initial evidence $B_0 = \{d_0\}$. The corresponding
amount of uncertainty is:

$$H(V_0) = 2.4355.$$

Step 1: d_1: "It is well-known that Jane wanted her husband dead and she
was seen at the place of the murder." Table 2 summarizes the list of suspects
given the body of evidence $B_1 = \{d_0, d_1\}$. The corresponding uncertainty is:

$$H(V_1) = 1.0889.$$

Step 2: d_2: "Jane attended the dinner party at Sir Montagu Corner at the time when Lord Edgware was murdered." Table 3 summarizes the list of suspects given the body of evidence $B_2 = \{d_0, d_1, d_2\}$. The corresponding uncertainty is:

$$H(V_2) = 0.69315.$$

Step 3: d_3: "Carlotta dies and the letter sent to her sister incriminates Ronald." Table 4 summarizes the list of suspects given the body of evidence $B_3 = \{d_0, d_1, d_2, d_3\}$. The corresponding uncertainty is:

$$H(V_3) = 1.3297.$$

Step 4: d_4: "Poirot discovers that a page was ingeniously removed from Carlotta's letter which now may incriminate any men not only Ronald." Table 5 summarizes the list of suspects given the body of evidence $B_4 = \{d_0, d_1, d_2, d_3, d_4\}$. The corresponding uncertainty is:

$$H(V_4) = 1.9459.$$

Step 5: d_5: "Poirot discovers that in Carlotta's letter the word *he* should be read *she* in which case the letter incriminates women not men." Table 6 summarizes the list of suspects given the body of evidence $B_5 = \{d_0, d_1, d_2, d_3, d_4, d_5\}$. The corresponding uncertainty is:

$$H(V_5) = 1.6094.$$

Step 6: d_6: "Jane's humiliating gaffe about the judgment of Paris ('Paris?' she said. 'Why, Paris doesn't cut any ice nowadays. It's London and New York that count.'), at Widburns' luncheon party at Claridge's, destroys evidence d_2, revealing that the sophisticated 'Jane', who delighted the guests at the dinner party at Sir Montagu Corner with her deep knowledge about the Greek mithology, was not the real Jane, but Carlotta, the impersonator." Table 7 summarizes the list of suspects given the body of evidence $B_6 = \{d_0, d_1, d_2, d_3, d_4, d_5, d_6\}$. The corresponding amount of uncertainty is:

$$H(V_6) = 0.$$

And Jane Wilkinson is arrested. She confesses murdering Lord Edgware.

V_0	motive	opportunity	$u_0(h, B_0)$	$\varphi_0\{p\}$	Bel_0	Pl_0	Amb_0
$\{w\}$	1	1	2	2/17	2/17	6/17	4/17
$\{c\}$	0	1	1	1/17	1/17	1/17	0
$\{m\}$	1	0	1	1/17	1/17	3/17	2/17
$\{r\}$	1	1	2	2/17	2/17	4/17	2/17
$\{s\}$	0	1	1	1/17	1/17	1/17	0
$\{g\}$	1	1	2	2/17	2/17	4/17	2/17
$\{a\}$	1	0	1	1/17	1/17	3/17	2/17
$\{b\}$	0	1	1	1/17	1/17	1/17	0
$\{w, m\}$	1	1/2	1.5	3/34	9/34	9/34	0
$\{g, r\}$	1	1	2	2/17	6/17	6/17	0
$\{w, a\}$	1	1/2	1.5	3/34	9/34	9/34	0
$\{c, m\}$	1/2	1/2	1	1/17	3/17	3/17	0

Table 1: Step 0 of the investigation.

V_1	motive	opportunity	$u_1(h, B_1)$	$\varphi_1\{p\}$	Bel_1	Pl_1	Amb_1
$\{w\}$	1	1	2	2/5	2/5	1	3/5
$\{w, m\}$	1	1/2	1.5	3/10	7/10	9/10	1/5
$\{w, a\}$	1	1/2	1.5	3/10	7/10	9/10	1/5

Table 2: Step 1 of the investigation.

V_2	motive	opportunity	$u_2(h, B_2)$	$\varphi_2\{p\}$	Bel_2	Pl_2	Amb_2
$\{c\}$	0	1	1	1/2	1/2	1	1/2
$\{c, m\}$	1/2	1/2	1	1/2	1	1	0

Table 3: Step 2 of the investigation.

V_3	*motive*	*opportunity*	$u_3(h, B_3)$	$\varphi_3\{p\}$	Bel_3	Pl_3	Amb_3
$\{c\}$	0	1	1	1/6	1/6	1/3	1/6
$\{r\}$	1	1	2	1/3	1/3	2/3	1/3
$\{g,r\}$	1	1	2	1/3	1/2	1/2	0
$\{c,m\}$	1/2	1/2	1	1/6	1/2	1/2	0

Table 4: Step 3 of the investigation.

V_4	*motive*	*opportunity*	$u_4(h, B_4)$	$\varphi_4\{p\}$	Bel_4	Pl_4	Amb_4
$\{c\}$	0	1	1	1/9	1/9	2/9	1/9
$\{m\}$	1	0	1	1/9	1/9	2/9	1/9
$\{r\}$	1	1	2	2/9	2/9	1/3	1/9
$\{a\}$	1	0	1	1/9	1/9	1/9	0
$\{b\}$	0	1	1	1/9	1/9	1/9	0
$\{g,r\}$	1	1	2	2/9	4/9	4/9	0
$\{c,m\}$	1/2	1/2	1	1/9	1/3	1/3	0

Table 5: Step 4 of the investigation.

V_5	*motive*	*opportunity*	$u_5(h, B_5)$	$\varphi_5\{p\}$	Bel_5	Pl_5	Amb_5
$\{c\}$	0	1	1	1/7	1/7	2/7	1/7
$\{s\}$	0	1	1	1/7	1/7	1/7	0
$\{g\}$	1	1	2	2/7	2/7	4/7	2/7
$\{g,r\}$	1	1	2	2/7	4/7	4/7	0
$\{c,m\}$	1/2	1/2	1	1/7	2/7	2/7	0

Table 6: Step 5 of the investigation.

V_6	*motive*	*opportunity*	$u_6(h, B_6)$	$\varphi_6\{p\}$	Bel_6	Pl_6	Amb_6
$\{w\}$	1	1	2	1	1	1	0

Table 7: Step 6 of the investigation.

Remark: Weighting probabilities seems to be an important characteristic of any intelligence-based system. The first technique of weighting probabilities was introduced, long ago, by Thomas Bayes (1702-1761), the English minister whose seminal paper *An essay towards solving a problem in the doctrine of chances* contained the formula for calculating the posterior, or retrodiction, probability $p(h \mid e)$ of a hypothesis h given the evidence e from the prior probability $p(h)$ of the hypothesis h and the prediction probability $p(e \mid h)$ of the evidence e given the hypothesis h, which, as formulated by Laplace (1812), has the expression:

$$p(h \mid e) = \frac{p(h)\,p(e \mid h)}{\sum_h p(h)\,p(e \mid h)}.$$

Apparently, Bayes was affraid to publish the above memntioned paper because applying the above formula meant to impose a special kind of evaluation of the new credibilities of available hypotheses based on the evidence at hand. Fortunately, after Bayes' death, Richard Price, a friend of his, found the paper in a drawer and sent it to the Royal Society in London, for publication. Two and a half centuries later, the Bayesian viewpoint for calculating the posterior probabilities, or credibilities, of the available hypotheses based on the new found evidence is still a controversial topic, but large parts of modern statistical inference and decision theory are essentially based on it. It is very difficult to think how pattern-recognition and learning theory could be managed without using somehow the Bayes formula and its many generalizations.

Appendix. REACHING A VERDICT

BY WEIGHTING EVIDENCE

1. Introduction. Subjective logic is very often used in everyday life. When the available evidence is only partial or even contradictory and is provided by not entirely reliable sources, making up our minds about what is going on involves an inevitable weighting of this evidence. The weighting process could be based either on additional information about the reliability of the available sources of evidence or could rely on similar cases from past experiences. Being often subjective, the weighting process of evidence can yield either correct or incorrect conclusions. Detective stories are so popular just because they aim at finding the truth behind a web of partial, incomplete, and often contradictory facts, alibis and testimonies. An investigator relies on the evidence at hand but is free to make correlations and suppositions that could reveal a new interpretation of the existing evidence or show a new direction where to search for new evidence before reaching a final conclusion. In doing this, the detective uses not only the available evidence but also his intuition, feelings, and analogies from his or other people's past experience. In his more serious pursuit, a scientific researcher is not so far from what an intelligent detective is doing.

The objective of this appendix (Guiasu, (1994b)) is to discuss a mathematical model dealing with a set X of potential suspects, one or several bodies of evidence provided by not entirely relevant facts collected by investigators or by testimonies made by not always entirely reliable witnesses, and a jury or a judge trying to reach a verdict by weighting the available evidence. A body of evidence induces a probability (or credibility) distribution on the class $\mathcal{P}(X)$ of all possible subsets of X. A judge is associated to a family of conditional weights that are nonnegative functions defined on $\mathcal{P}(X)$ conditioned by the given evidence. The weights and bodies of evidence are combined for getting a weighted probability (credibility) distribution on $\mathcal{P}(X)$ which is used for reaching a verdict about the culpability or innocence of the subsets of the universe X. An example from an Agatha Christie's mystery short story is given both for illustrating the significance of the mathematical symbols introduced and for underlying the difference between objective and subjective weightings of the same body of evidence. The main part of this appendix shows how the classic decision making rules formulated by Hooper, Dempster, Bayes, and Jeffrey are special cases of weighting bodies of evidence. The case of a body of evidence induced by a fuzzy set is also discussed.

2. The model involving direct evidence. The present mathematical model contains a finite set of suspects called universe, bodies of evidence inducing probability (credibility) distributions on the class of all subsets of the universe, and a judge associated with a family of conditional weights which is a family of nonnegative functions on $\mathcal{P}(X)$ conditioned by the available evidence. The bodies of evidence and the weights are combined in order to get a weighted probability (credibility) distribution on $\mathcal{P}(X)$ which may be used for calculating the belief, plausibility, and ambiguity of the subsets of the universe X and reach a verdict about the culpability or innocence of the potential suspects.

A. The universe. Let X be a crisp (Cantor) finite set called universe or frame. Let $\mathcal{P}(X)$ be the class of all subsets of X. Intuitively, we may think of X as being a set of potential suspects in a criminal investigation in which case $\mathcal{P}(X)$ is the class of all possible subsets of suspects, while the empty set $\emptyset$ means that there is no suspect in X.

B. Bodies of evidence. The evidence is provided either by testimonies made by witnesses or by facts collected by investigators on the culpability or innocence of subsets of the universe X.

(a) *Simple evidence.* Simple evidence refers to the case when the bodies of evidence are mutually independent. A body of evidence induces a probability (credibility) distribution on $\mathcal{P}(X)$. Thus $m_i : \mathcal{P}(X) \longrightarrow [0,1]$ and:

$$m_i(A) \geq 0, \quad (A \in \mathcal{P}(X)), \quad \sum_{A \subseteq X} m_i(A) = 1.$$

The number $m_i(A)$ denotes the probability (or credibility) that the suspects belong to the subset A but not to a subsubset of it. The class of *focal* subsets of X corresponding to m_i is:

$$\mathcal{F}(X; m_i) = \{A; A \subseteq X, m_i(A) > 0\}.$$

The *belief, plausibility* and *ambiguity* of A induced by m_i are defined by:

$$Bel(A; m_i) = \sum_{B \subseteq A, B \neq \emptyset} m_i(B),$$

$$Pl(A; m_i) = \sum_{B \cap A \neq \emptyset} m_i(B),$$

$$Amb(A; m_i) = \sum_{B \cap A \neq \emptyset, B \not\subseteq A} m_i(B).$$

If $\bar{C} = X - C$ is the complementary subset of C, then we obviously have:

$$m_i(\emptyset) + Bel(C; m_i) + Pl(\bar{C}; m_i) = 1,$$

$$Bel(C; m_i) \leq Pl(C; m_i).$$

The ambiguity of A induced by m_i, namely $Amb(A; m_i)$, takes into account only the subsets B of the universe X that make both A and its complement $\bar{A}$ plausible, i.e., it is obtained by summing up the values of m_i for all the subsets of X except the proper subsets of A and the proper subsets of $\bar{A}$.

(b) *Mixed evidence.* A pair of dependent bodies of evidence, let us say witness i and witness j testifying dependently, induce a joint probability (credibility) distribution, namely:

$$m_{ij} : \mathcal{P}(X) \times \mathcal{P}(X) \longrightarrow [0, 1],$$

$$m_{ij}(A, B) \geq 0, \quad \sum_{A \subseteq X} \sum_{B \subseteq X} m_{ij}(A, B) = 1,$$

where $m_{ij}(A, B)$ is the probability (credibility) that witness i focuses on subset A and witness j focuses on subset B. If the bodies of evidence are independent, then:

$$m_{ij}(A, B) = m_i(A) \, m_j(B).$$

If $m_j(B) > 0$, the conditional probability (credibility) distribution on $\mathcal{P}(X)$ given B is:

$$m_{i|j}(A \mid B) = m_{ij}(A, B)/m_j(B).$$

The corresponding class of focal pairs of subsets is:

$$\mathcal{F}(X, X; m_{ij}) = \{(A, B); A \subseteq X, B \subseteq X, m_{ij}(A, B) > 0\}.$$

In a natural way we can introduce the functions $BelBel$, $BelPl$, $PlPl$, $BelAmb$, etc., on $\mathcal{P}(X) \times \mathcal{P}(X)$. Thus, for instance,

$$BelPl(A, B; m_{ij}) = \sum_{C \subseteq A, C \neq \emptyset} \sum_{D \cap B \neq \emptyset} m_{ij}(C, D).$$

Obviously, if the bodies of evidence i and j are independent, then $BelPl$ is equal to $Bel \times Pl$, or:

$$BelPl(A, B; m_{ij}) = Bel(A; m_i)\, Pl(B; m_j).$$

C. The judge. A judge, or decision maker, or jury, has to reach a verdict about the culpability or innocence of the suspects based on the available evidence and his own judgement. Mathematically, the judge is associated with a family of conditional weights which is a family of nonnegative functions on $\mathcal{P}(X)$ conditioned by the available evidence. Thus, the weights corresponding to the body of evidence #i for which m_i is the probability (credibility) distribution induced on $\mathcal{P}(X)$ are:

$$w_i(\cdot \mid \cdot) : \mathcal{P}(X) \times \mathcal{F}(X; m_i) \longrightarrow [0, \infty),$$

where $w_i(C \mid A)$ represents judge's weight assigned to the culpability of the subset $C \in \mathcal{P}(X)$ if the i-th body of evidence focuses on the culpability of the subset $A \in \mathcal{F}(X; m_i)$. Now, the i-th body of evidence weighted by the judge will provide the new probability (credibility) distribution on $\mathcal{P}(X)$ given by:

$$\mu_i(C) = \sum_{A \in \mathcal{F}(X; m_i)} w_i(C \mid A)\, m_i(A), \tag{1}$$

abbreviated by $\mu_i = w_i \star m_i$. The larger the weight the larger the potential culpability of the corresponding subset of suspects. From mathematical point of view, except nonnegativity, the only condition imposed on the family of weights is:

$$\sum_{C \in \mathcal{P}(X)} \sum_{A \in \mathcal{F}(X; m_i)} w_i(C \mid A)\, m_i(A) = 1, \tag{2}$$

which implies that μ_i given by (1) is a probability (credibility) distribution on $\mathcal{P}(X)$. The corresponding class of focal subsets $\mathcal{F}(X; \mu_i)$, belief function $Bel(\cdot; \mu_i)$, and plausibility function $Pl(\cdot; \mu_i)$ may be defined in the usual way.

A family of weights is *probabilistic* if they satisfy the equalities:

$$\sum_{C \in \mathcal{P}(X)} w_i(C \mid A) = 1, \qquad \text{for every} \quad A \in \mathcal{F}(X; m_i). \tag{3}$$

In such a case, very often the weights represent the judge's credibility in the culpability of the subsets of suspects conditioned by the evidence available. Obviously, (3) implies (2) but the converse is not necessarily true. If the family of weights is probabilistic and objective, based exclusively on relative frequencies, then $w_i(C \mid A)$ may be calculated using the standard formula for conditional probabilities. If, however, the family of weights is both non-probabilistic and subjective, then $w_i(C \mid A)$ simply reflects what the judge believes about the culpability of C if the direct evidence focuses on the subset A and no special rule is necessarily used for getting it.

Particular cases: (a) If the judge fully relies on the i-th body of evidence, then $w_i(A \mid A) = 1$ and $w_i(C \mid A) = 0$ if C is different from A, for every $A \in \mathcal{F}(X; m_i)$, which implies $\mu_i(A) = m_i(A)$.

(b) If the judge focuses on $B \in \mathcal{P}(X)$ regardless of what the i-th body of evidence says, then $w_i(B \mid A) = 1$ for every $A \in \mathcal{F}(X; m_i)$, which implies $\mu_i(B) - 1$.

In general, a weight w is *fully compatible* with the probability (credibility) distribution m induced by a body of evidence if $w(A \mid A) = 1$ and $w(C \mid A) = 0$ if C is different from A, for every $A \in \mathcal{F}(X; m)$. Also, w is *compatible* with m if $w(B \mid A) = 0$ for every $B \notin \mathcal{F}(X; m)$. A weight is not fully compatible with m if either the corresponding body of evidence is not reliable or the judge is biased against or in favour of that evidence. A weight is fully compatible with m if the judge gives full credit to the body of evidence that induces m on $\mathcal{P}(X)$. The weighted body of evidence is *conclusive* only if there is a subset $A \in \mathcal{P}(X)$ such that $\mu(A) = 1$.

Let us notice that if $\mathcal{F}(X; \mu_i) - \mathcal{F}(X; m_i)$ and $[w_i(\cdot \mid \cdot)]$ is a doubly stochastic matrix, then the entropy of μ_i is larger than or equal to the entropy of m_i, which shows that there is more confusion about what subset of X to focus on after weighting the available evidence than before. Maximum confusion is attained when $w_i(C \mid A) = 1/s_i$ for every $A \in \mathcal{F}(X; m_i)$, where s_i is the number of subsets of the class $\mathcal{F}(X; m_i)$, in which case the entropy of μ_i is equal to $\log s_i$. More often than not, however, the weighting process diminishes the uncertainty on the class of possible subsets of X, and sometimes even drastically.

All the considerations made above may be formulated in a straightforward way when we assign weights $w_{ij}(\cdot \mid \cdot, \cdot)$ to a mixed evidence inducing the joint probability (credibility) distribution m_{ij} on $\mathcal{P}(X) \times \mathcal{P}(X)$. Thus, for instance, the probability (credibility) distribution induced on $\mathcal{P}(X)$ by the weighted mixed (i, j)-th body of evidence is:

$$\mu_{ij}(C) = \sum_{(A,B) \in \mathcal{F}(X,X;m_{ij})} w_{ij}(C \mid A, B)\, m_{ij}(A, B), \quad (C \in \mathcal{P}(X)), \quad (4)$$

where $w_{i,j}(C \mid A, B)$ is the judge's weight of the subset $C \in \mathcal{P}(X)$ given the mixed evidence $(A, B) \in \mathcal{F}(X, X; m_{i,j})$.

Remarks: (a) The idea of associating a body of evidence to a probability distribution m on $\mathcal{P}(X)$ is due to Dempster (1967) and extensively discussed in Shafer (1976). In their approach, however, $m(\emptyset)$ has to be always equal to zero, an unnecessary restriction because m is not obtained by extending a probability distribution on X to a probability measure on $\mathcal{P}(X)$, but is directly defined as a probability distribution on $\mathcal{P}(X)$, in which case $m(\emptyset)$ could be positive, corresponding to the frequent case when there is a positive probability of having nobody guilty in the universe X.

(b) Let $\chi_F : X \longrightarrow [0, 1]$ be a fuzzy set in Zadeh's sense (Zadeh, (1965)). Then, the number $\chi_F(x)$ is the degree of membership of the element $x \in X$ to the fuzzy set F. Obviously, $\{\chi_F(x); x \in X\}$ is not a probability distribution on X but it induces a probability distribution m_F on $\mathcal{P}(X)$, defined by:

$$m_F(A) = \prod_{x \in A} \chi_F(x) \prod_{y \in \bar{A}} [1 - \chi_F(y)], \quad A \in \mathcal{P}(X),$$

describing the body of evidence induced by a fuzzy set F. Let us notice that if $0 < \chi_F(x) < 1$, for every $x \in X$, then $\mathcal{F}(X; m) = \mathcal{P}(X)$. All the considerations made above could be applied to the case when the available evidence is provided by fuzzy sets defined on X.

3. The model involving indirect evidence. There are cases when the judge has to reach a verdict using indirect evidence. This more general case may be treated like the previous one with some minor changes. Thus, let X, Y, and Z be three finite crisp (Cantor) sets, where X is the judge's universe and Y, Z are universes of two bodies of evidence.

(a) *Simple evidence.* If a body of evidence induces a probability (credibility) distribution m on $\mathcal{P}(Y)$ and the judge's weights are:

$$w(\cdot \mid \cdot) : \mathcal{P}(X) \times \mathcal{F}(Y; m) \longrightarrow [0, \infty),$$

then the weighted probability (credibility) distribution on $\mathcal{P}(X)$ will be:

$$\mu(C) = \sum_{A \in \mathcal{F}(Y;m)} w(C \mid A)\, m(A),$$

for every $C \in \mathcal{P}(X)$, where:

$$\sum_{C \in \mathcal{P}(X)} \sum_{A \in \mathcal{F}(Y;m)} w(C \mid A)\, m(A) = 1.$$

(b) *Mixed evidence.* If two dependent bodies of evidence, say #1 and #2, induce a joint probability (credibility) distribution $m_{1,2}$ on $\mathcal{P}(Y) \times \mathcal{P}(Z)$, and $w_{1,2}(\cdot \mid \cdot, \cdot)$ is the family of judge's weights on $\mathcal{P}(X)$ given the mixed evidence from $\mathcal{F}(Y, Z; m_{1,2})$, then the weighted probability (credibility) distribution on $\mathcal{P}(X)$ is:

$$\mu_{1,2}(C) = \sum_{(A,B) \in \mathcal{F}(Y,Z;m_{1,2})} w_{1,2}(C \mid A, B)\, m_{1,2}(A, B), \tag{5}$$

where:

$$\sum_{C \in \mathcal{P}(X)} \sum_{(A,B) \in \mathcal{F}(Y,Z;m_{1,2})} w_{1,2}(C \mid A, B)\, m_{1,2}(A, B) = 1,$$

for every $(A, B) \in \mathcal{F}(Y, Z; m_{1,2})$.

4. An example. All probability distributions involved in the above analysis could be either objective, i.e., based on relative frequencies of the outcomes of probabilistic experiments repeated independently, or subjective, in which case they are rather called credibility distributions. The weights themselves can be either objective or subjective. The subjective weights can be based on judge's past experience, logic, feelings, and/or imagination. The result of the weighting process could be a right or wrong verdict, expressed in a sharp or vague form. An Agatha Christie's mystery short story, namely *The Tuesday Night Club* (Christie, (1984b, pp.1-14)), offers a very good example of different kinds of evidence and weights that could be used in reaching correct, partially correct, or incorrect verdicts.

Briefly, Mrs. Jones has died and the characters are: J (Mr. Jones, i.e., Mrs. Jones's husband), C (Miss Clark, Mrs. Jones's companion), Dr (the doctor), D (the doctor's daughter), and G (Gladys Linch, the maid). Thus, the universe is $X = \{J, C, Dr, D, G\}$. In what follows, the body of evidence #i implies ($\Rightarrow$) the credibility distribution $m_i : \mathcal{P}(X) \longrightarrow [0, 1]$, where $m_i(\{C, D\})$, for instance, means the credibility that C and D, together, have murdered Mrs. Jones, as a possible result of the body of evidence #i. Also, $m_i(\emptyset)$ means the credibility induced by the body of evidence #i that nobody is guilty and the death has occurred as an unfortunate accident. Thus, $m_i(\emptyset) = 1$ means that the body of evidence #i does not incriminate anybody. Also, only the positive weights are going to be explicitly mentioned.

Body of evidence #1: J, C, and Mrs. Jones sat down to a supper consisting of tinned lobster and salad, trifle, and bread and cheese. Later in the night all three were taken ill, and a doctor was hastily summoned. Two people recovered, Mrs. Jones died. Death was considered to be due to ptomaine poisoning, a certificate was given to that effect and the victim was duly buried. $\Rightarrow m_1(\emptyset) = 1$.

Body of evidence #2: J had been staying the previous night at a small hotel in Birmingham and the chambermaid there found on the blotting paper the text: "Entirely dependent on my wife ... when she is dead I will ... hundreads and thousands ..." Also, J had been very attentive to D. He also benefited by his wife's death to the amount of £8000. $\Rightarrow m_2(\{J\}) = 1$.

Body of evidence #3: An exhumation was ordered. The result of the autopsy was that the deceased lady had died of arsenical poisoning. $\Rightarrow$

$$m_3(\{J\}) = m_3(\{C\}) = m_3(\{Dr\}) = m_3(\{D\}) = m_3(\{G\}) = 1/5.$$

Body of evidence #4: J's testimony. (The friendship with D had been over two months before the death; in Birmingham he wrote in fact an innocent letter to his brother; he returned from Birmingham just as supper was being served.) $\Rightarrow m_4(\emptyset) = 1$.

Body of evidence #5: Dr's testimony and the investigation on him made by Scotland Yard. $\Rightarrow m_5(\emptyset) = 1$.

Body of evidence #6: After supper, J had gone down to the kitchen and had demanded a bowl of corn-flour for his wife prepared by G. He had waited in the kitchen and then carried it up to his wife's room himself. J had motive and opportunity to kill his wife. $\Rightarrow m_6(\{J\}) = 1$.

Body of evidence #7: C's testimony. (The whole of the bowl of corn-flour was drunk by her. As she was banting at the time, she was always hungry and Mrs. Jones had changed her mind about tasting the corn-flour.) $\Rightarrow m_7(\emptyset) = 1$.

At this moment, Sir Henry, a commissioner of Scotland Yard, asks Joyce, Mr. Petherick, Raymond, and Miss Marple to reach individually a verdict based on the seven independent bodies of evidence. In computation, we are using here the formula (4) with:

$$m_{1,2,3,4,5,6,7} = m_1\,m_2\,m_3\,m_4\,m_5\,m_6\,m_7.$$

Their conclusions are the following:

Mr. Petherick (a solicitor, relying on facts and money): J was guilty and C sheltered him for money, lying about drinking the corn-flour.

$$\Rightarrow w(\{J,C\} \mid \emptyset, \{J\}, A_3, \emptyset, \emptyset, \{J\}, \emptyset) = 1,$$

where, here and subsequently, A_3 has to be successively replaced by the possible subsets induced by the body of evidence #3, namely $\{J\}$, $\{C\}$, $\{Dr\}$, $\{D\}$, and $\{G\}$. Applying (4), we get the verdict $\mu(\{J,C\}) = 1$.

Joyce (a young artist, relying on intuition): C was guilty because probably she was in love with J and hated his wife.

$$\Rightarrow w(\{C\} \mid \emptyset, \{J\}, A_3, \emptyset, \emptyset, \{J\}, \emptyset) = 1, \quad \Rightarrow \mu(\{C\}) = 1.$$

Raymond (a young writer, relying on imagination): D was guilty. After noticing the poisoning symptoms, Dr sent a messenger home for some opium pills for Mrs. Jones to relieve her acute pain. D, who was in love with J, had motive and opportunity. Consequently, she sent back pills containing pure white arsenic.

$$\Rightarrow w(\{D\} \mid \emptyset, \{J\}, A_3, \emptyset, \emptyset, \{J\}, \emptyset) = 1, \quad \Rightarrow \mu(\{D\}) = 1.$$

Miss Marple (a smart old lady, relying on life experience and analogy): A similar case happened in the village St. Mary Mead. G murdered Mrs. Jones, pushed by J who made her his murder instrument.

$$\Rightarrow w(\{G, J\} \mid \emptyset, \{J\}, A_3, \emptyset, \emptyset, \{J\}, \emptyset) = 1, \quad \Rightarrow \mu(\{G, J\}) = 1.$$

Scotland Yard (objective weighting, giving credit only to the available evidence): Taking into account only the seven bodies of evidence and calculating how many times each suspect has been involved, we get:

$$w(\emptyset \mid \emptyset, \{J\}, \{J\}, \emptyset, \emptyset, \{J\}, \emptyset) = \frac{4}{7},$$

$$w(\{J\} \mid \emptyset, \{J\}, \{J\}, \emptyset, \emptyset, \{J\}, \emptyset) = \frac{3}{7},$$

$$w(\emptyset \mid \emptyset, \{J\}, \{C\}, \emptyset, \emptyset, \{J\}, \emptyset) = \frac{4}{7},$$

$$w(\{J\} \mid \emptyset, \{J\}, \{C\}, \emptyset, \emptyset, \{J\}, \emptyset) = \frac{2}{7},$$

$$w(\{C\} \mid \emptyset, \{J\}, \{C\}, \emptyset, \emptyset, \{J\}, \emptyset) = \frac{1}{7},$$

$$w(\emptyset \mid \emptyset, \{J\}, \{Dr\}, \emptyset, \emptyset, \{J\}, \emptyset) = \frac{4}{7},$$

$$w(\{J\} \mid \emptyset, \{J\}, \{Dr\}, \emptyset, \emptyset, \{J\}, \emptyset) = \frac{2}{7},$$

$$w(\{Dr\} \mid \emptyset, \{J\}, \{Dr\}, \emptyset, \emptyset, \{J\}, \emptyset) = \frac{1}{7},$$

$$w(\emptyset \mid \emptyset, \{J\}, \{D\}, \emptyset, \emptyset, \{J\}, \emptyset) = \frac{4}{7},$$

$$w(\{J\} \mid \emptyset, \{J\}, \{D\}, \emptyset, \emptyset, \{J\}, \emptyset) = \frac{2}{7},$$

$$w(\{D\} \mid \emptyset, \{J\}, \{D\}, \emptyset, \emptyset, \{J\}, \emptyset) = \frac{1}{7},$$

$$w(\emptyset \mid \emptyset, \{J\}, \{G\}, \emptyset, \emptyset, \{J\}, \emptyset) = \frac{4}{7},$$

$$w(\{J\} \mid \emptyset, \{J\}, \{G\}, \emptyset, \emptyset, \{J\}, \emptyset) = \frac{2}{7},$$

$$w(\{G\} \mid \emptyset, \{J\}, \{G\}, \emptyset, \emptyset, \{J\}, \emptyset) = \frac{1}{7}.$$

No evidence pointed at subsets with more than one suspect. From (4), we get the following values:

$$\mu(\emptyset) = \tfrac{4}{7}; \quad \mu(\{J\}) = \tfrac{11}{35}; \quad \mu(\{C\}) = \tfrac{1}{35};$$
$$\mu(\{Dr\}) = \tfrac{1}{35}; \quad \mu(\{D\}) = \tfrac{1}{35}; \quad \mu(\{G\}) = \tfrac{1}{35}.$$

Consequently, although there was a high credibility $(11/35)$ in J's culpability, the evidence was not considered conclusive and no arrest was made.

Body of evidence #8: One year later, G's testimony before dying. (J promised to marry her when his wife was dead. Following J's instructions, she put arsenic in trifle. Only Mrs. Jones ate it because C was on diet and J knew about the poison. She had a child with J. The child died at birth and J deserted her for another woman.) $\Rightarrow m_8(\{G, J\}) = 1$.

Scotland Yard (based on this final evidence):

$$\Rightarrow w(\{G, J\} \mid \{G, J\}) = 1, \Rightarrow \mu(\{G, J\}) = 1.$$

5. Special case: Hooper's rule. Let us take two independent bodies of evidence provided by two witnesses inducing the probability distributions m_1 and m_2 on $\mathcal{P}(X)$, respectively. Let:

$$\mathcal{F}(X; m_1) = \mathcal{F}(X; m_2) = \{A, \bar{A}\},$$

where $\bar{A} = X - A$, which shows that both witnesses focus on the same subsets A and $\bar{A}$ of X. The weights are defined as:

$$w_{1,2}(A \mid A, A) = w_{1,2}(A \mid A, \bar{A}) =$$

$$= w_{1,2}(A \mid \bar{A}, A) = w_{1,2}(\bar{A} \mid \bar{A}, \bar{A}) = 1,$$

which means that the judge gives full credit to A if at least one witness focuses on A and full credit to $\bar{A}$ only if both witnesses focus on $\bar{A}$. Then, according to (1), we have:

$$\mu_{1,2}(\bar{A}) = m_1(\bar{A}) \, m_2(\bar{A}), \quad \mu_{1,2}(A) = 1 - [1 - m_1(A)][1 - m_2(A)].$$

According to Lindley (1987a), this rule for combining evidence was used by G. Hooper in 1685.

6. Special case: Dempster's rule. Let us take two independent bodies of evidence provided by two witnesses inducing the probability (credibility) distributions m_1 and m_2 on $\mathcal{P}(X)$. The judge takes into account only the common part of the focal sets of the two witnesses and the only positive weights are:

$$w_{1,2}(A \cap B \mid A, B) = \left[1 - \sum_{C \in \mathcal{F}(X; m_1), D \in \mathcal{F}(X; m_2), C \cap D = \emptyset} m_1(C) \, m_2(D) \right]^{-1},$$

for all $A \in \mathcal{F}(X; m_1), B \in \mathcal{F}(X; m_2), A \cap B \neq \emptyset$. In such a case (4) becomes:

$$\mu_{1,2}(C) = \left[\sum_{C = A \cap B} m_1(A) \, m_2(B) \right] \bigg/ \left[1 - \sum_{A \cap B = \emptyset} m_1(A) \, m_2(B) \right],$$

which is Dempster's rule (Dempster, (1967)) of combining two independent bodies of evidence. It gives equal credit to the common evidence and discards any other evidence. According to Shafer (1976), in the special case of a universe containing only two elements, this rule was used by J.H. Lambert in his Neues Organon published in 1764.

7. Special case: Jeffrey's rule. *First interpretation:* Let us take two bodies of evidence inducing the probability (credibility) distributions m_1 and m_2 on $\mathcal{P}(X)$, respectively, where $X = Y \times Z$ is the Cartesian product of two finite crisp (Cantor) sets. Assume:

$$\mathcal{F}(X; m_1) = \{\{(y, z); z \in Z\}; y \in Y\}, \quad m_1(\{(y, z); z \in Z\}) = p_Y(y);$$

$$\mathcal{F}(X; m_2) = \{\{(y, z); y \in Y\}; z \in Z\}, \quad m_2(\{(y, z); y \in Y\}) = q_Z(z),$$

where p_Y and q_Z are probability distributions on Y and Z, respectively. Let $p_Z(\cdot \mid y)$ be a conditional probability distribution on Z given $y \in Y$, and p_Z the *prediction* probability distribution on Z defined by:

$$p_Z(z) = \sum_{y \in Y} p_Z(z \mid y) p_Y(y),$$

where p_Y is interpreted as being the *prior* probability distribution on Y. Let:

$$\mu_{1,2}(C) = \sum_{A \in \mathcal{F}(X; m_1)} \sum_{B \in \mathcal{F}(X; m_2)} w(C \mid A, B) \, m_1(A) \, m_2(B),$$

where:

$$\mathcal{F}(X;\mu_{1,2}) = \{\{(y,z)\}; y \in Y, z \in Z\} = \{A \cap B; A \in \mathcal{F}(X;m_1), B \in \mathcal{F}(X;m_2)\},$$

$$w(\{\{(y,z)\} \mid \{(y,z); z \in Z\}, \{(y,z); y \in Y\}) = p_Z(z \mid y)/p_Z(z).$$

Then, we have:

$$\mu_{1,2}(\{(y,z)\}) = \frac{p_Y(y)\,p_Z(z \mid y)}{p_Z(z)}\,q_Z(z),$$

which is a probability distribution on $Y \times Z$. Its marginal probability distribution, namely,

$$
\begin{aligned}
p_Y(y \mid q_Z) &= Bel(\{(y,z); z \in Z\}; \mu_{1,2}) = \sum_{z \in Z} \mu_{1,2}(\{(y,z)\}) \\
&= \sum_{z \in Z} \frac{p_Y(y)\,p_Z(z \mid y)}{p_Z(z)}\,q_Z(z),
\end{aligned}
\tag{6}
$$

is Jeffrey's rule (Jeffrey, (1965); Ichihashi and Tanaka, (1989)), for calculating the *posterior* probability distribution on Y given the *actual* probability distribution q_Z on Z.

Second interpretation: Jeffrey's rule may be obtained more directly from (5), as a weighting with indirect evidence. Indeed, let X and Y be two finite crisp (Cantor) sets and m a probability distribution on $\mathcal{P}(Y)$ such that:

$$\mathcal{F}(Y;m) = \{\{y\}; y \in Y\}, \quad m(\{y\}) = q(y),$$

where q is the *actual* probability distribution on Y. Taking the only positive weights on $\mathcal{P}(X)$ to be:

$$w(\{x\} \mid \{y\}) = \frac{p(y \mid x)\,p(x)}{\sum_{x \in X} p(y \mid x)\,p(x)}$$

where p is a *prior* probability distribution on X and $p(\cdot \mid x)$ is a conditional probability distribution on Y given $x \in X$, the weighted probability distribution (5) becomes:

$$
\begin{aligned}
p(x \mid q) &= \mu(\{x\}) = \sum_{y \in Y} w(\{x\} \mid \{y\})\,q(y) = \\
&= \sum_{y \in Y} \frac{p(y \mid x)\,p(x)}{\sum_{x \in X} p(y \mid x)\,p(x)}\,q(y),
\end{aligned}
\tag{7}
$$

which is Jeffrey's rule for calculating the *posterior* probability distribution on X. In this case,

$$\mathcal{F}(X;\mu) = \{\{x\}; x \in X\}.$$

8. Special case: Bayes' rule. Taking:

$$q_Z(z) = \begin{cases} 1, & \text{if } z = z_0; \\ 0, & \text{if } z \neq z_0, \end{cases}$$

the formula (6) becomes the Bayes' rule for calculating the *posterior* probability distribution, namely,

$$p_Y(y \mid z_0) = p_Y(y \mid q_Z) = \frac{p_Z(z_0 \mid y)\, p_Y(y)}{p_Z(z_0)}.$$

Bayes' rule may also be obtained from (7) by taking q to be a degenerate probability distribution focussed on a single element $\{y_0\}$, i.e., $q(\{y_0\}) = 1$.

9. Conclusions. The process of reaching a verdict essentially depends on how the available evidence is used by the judge or jury. The evidence may be significant, partially relevant, or misleading and the judge may use it in an objective or subjective way. The appendix discussed a general mathematical model of weighting the available evidence. The classic rules from decision theory proposed by Hooper, Dempster, Bayes, and Jeffrey prove to be special cases of such a weighting of evidence process. It is often said that only Bayes' rule is rational and the other alternatives are not. The appendix pleads in favour of a relaxed or relative rationality. Instead of the drastic dichotomy "rational or not", a mathematical model in decision theory should be classified as "rational with respect to a specific way of weighting the available evidence." It is often forgotten that Bayes' formula is a rigorous mathematical statement only when it is applied to the events of the same probability space. If this formula is used by analogy to describe the connection between different types of entities, like the available hypotheses, on one side, and the outcomes of an auxiliary experiment, on the other side, for instance, then it is only a viewpoint, one of the many possible ways of describing the respective interdependence. Perhaps Bayes realized this when he hesitated to publish his seminal paper (Bayes, (1763)) during his lifetime. Fortunately for us, as mentioned at the end of chapter 5, a close friend of his found the manuscript among his papers and published it posthumously.

References

[1] Ackermann, W. (1949). *Grundzüge der Theoretischen Logik.* Springer Verlag, Berlin.

[2] Aleksandrov, A.D., Kolmogorov, A.N., and Lavrentev, M.A. (1999). *Mathematics. Its Content, Methods, and Meaning.* Dover Publications, Mineola, New York.

[3] Allen, W. (1980). *Side Effects.* Ballantine Books, New York.

[4] Äquist, L. (1963). Postulate sets and decision procedure for some systems of deontic logic. *Teoria,* **29**, 154-175.

[5] Arnheim, R. (1971). *Entropy and Art. An Essay on Disorder and Order.* University of California Press, Berkeley and Los Angeles.

[6] Bar Hillel, Y. and Carnap, R. (1953). Semantic information. *The British Journal for the Philosophy of Science,* **4**, 145-157.

[7] Bargainnier, E.F. (1980). *The Gentle Art of Murder.* Bowling Green University Popular Press, Ohio.

[8] Barnard, R. (1980). *A Talent to Deceive.* Dodd, Mead & Company, New York.

[9] Bass, T.A. (1999). *The Predictors.* Henry Holt and Company, New York.

[10] Bayes, T.R. (1763). An essay towards solving a problem in the doctrine of chances. *Transactions of the Royal Society in London,* **53**, 370-418.

[11] Becker, O. (1930). Zur Logik der Modalitäten. *Jahrbuch für Philosophische und Phenomenologische Forschung,* **11**, 496-548.

[12] Belis, M. and Guiasu, S. (1968). A quantitative-qualitative measure of information in cybernetic systems. *IEEE Transactions of Information Theory,* **IT-14**, 593-594.

[13] Bell, E.T. (1965). *Men of Mathematics.* Simon & Schuster, Inc., New York.

[14] Betts, R.K. (1982). *Surprise Attack. Lesons for Defense Planning.* The Brookings Institution, Washington, D.C.

[15] Binkley, R. (1968). The surprise examination in modal logic. *Journal of Philosophy*, **65**, 127-136.

[16] Birkhoff, G. and von Neumann, J. (1936). The logic of quantum mechanics. *Annals of Mathematics*, **37**, 823-843.

[17] Bocivar, D.A. (1938). On the three-valued calculus and its applications to the analysis of paradoxes of the extended classic functional calculus. *Matematiceskii Sbornik*, **4**, 257-308.

[18] Bocivar, D.A. (1943). On the problem of noncontradiction of a three-valued calculus. *Matematiceskii Sbornik*, **12**, 353-369.

[19] Boole, G. (1958). *An Investigation of the Laws of Thought on which are Founded the Mathematical Theories of Logic and Probabilities.* Dover Publications, New York. (The first American printing of the work originally published by Macmillan in 1854).

[20] Boole, G. (1956). Mathematical analysis of logic. In Newman, J.R. (ed.) *The World of Mathematics.* Vol.3 (11th edition), Simon and Schuster, New York, pp.1856-1858.

[21] Borel, E. (1950). *Probabilité et certitude.* Presses Universitaires de France. Paris.

[22] Borkowski, L. (ed.) (1970). *Łukasiewicz, J. Selected Works.* North-Holland, Amsterdam.

[23] Boslough, J. (1985). *Stephen Hawking's Universe.* William Morrow and Company, New York.

[24] Botezatu, P., Dima, T., Bieltz, P., Vieru, S., and Enescu, Gh. (1974). *Direcţii în logica contemporană.* Editura Ştiinţifică, Bucureşti.

[25] Bouchon-Meunier, B. (1993). *La logique floue.* Presses Universitaires de France, Paris.

[26] Bourbaki, N. (1968). *Theory of Sets*. Hermann, Paris, Addison-Wesley Publishing Company, Reading, Massachusetts.

[27] Bourbaki, N. (1999). *Elements of the History of Mathematics*. Springer-Verlag, Berlin, Heidelberg, and New York.

[28] Brace, E.R. (1977). *An Illustrated Dictionary of Chess*. Hamlyn Publishing Group Lt., London.

[29] Broglie, L. de (1924). *Recherches sur la théorie des quanta*. Masson, Paris.

[30] Burke, J. (1978). *Connections*. Macmillan London Limited, London.

[31] Carnap, R. (1962). *Logical Foundations of Probability*. (2nd ed.) University of Chicago Press, Chicago.

[32] Castañeda, N. (1957). On the logic of norms. *Methodos*, **9**, 209-215.

[33] Casti, J.L. (1989). *Paradigms Lost*. Abacus, London, U.K.

[34] Caude, R. and Moles, A. (eds.) (1964). *Méthodologie. Vers une science de l'action*. Gauthier-Villars, Paris.

[35] Cavaillès, J. (1962). *Philosophie mathématique*. Hermann, Paris.

[36] Christie, A. (1984a). *Lord Edgware Dies*. Berkley Books, New York. (first published by Grosset & Dunlap in 1933).

[37] Christie, A. (1984b). *The Thirteen Problems*. Berkley Books, New York.

[38] Cooper, J.C. (1981). *Yin & Yang. The Taoist Harmony of Opposites*. The Aquarian Press, Wellingborough, Northamptonshire.

[39] Daniel, C. (ed.) (1987). *Chronicle of the 20th Century*. Chronicle Publications Inc., Mount Kisko, N.Y.

[40] De Luca, A. and Termini, S. (1972). A definition of a nonprobabilistic entropy in the setting of fuzzy sets. *Information and Control*, **20**, 301-312.

[41] De Luca, A. and Termini, S. (1979). Entropy and energy measures of a fuzzy set. In Gupta, M.M., Ragade, R.K., and Yager, R.R. (eds.) *Advances in Fuzzy Set Theory and Applications.* North-Holland Publishing Company, New York, pp.321-338.

[42] De Morgan, A. (1947). *Formal Logic.* Taylor and Walter, London.

[43] Dempster, A.P. (1967). Upper and lower probabilities induced by a multivalued mapping. *Annals of Mathematical Statistics,* **38**, 325-339.

[44] Desai, M.M. (1939). *Surprise. A Historical and Experimental Study.* University Press, Cambridge.

[45] Devine, B. and Cohen, J.E. (1992). *Absolute Zero Gravity.* Simon & Schuster Inc., New York.

[46] Dewey, R.E. and Gould, J.A. (1970). *Freedom. Its History, Nature, and Varieties.* Macmillan Publishing Co., New York.

[47] Dieudonné, J. (1992). *Mathematics – The Music of Reason.* Springer-Verlag, Berlin and Heidelberg

[48] Dima, T. (1975). *Metodele inductive.* Editura Ştiinţifică, Bucureşti.

[49] Dubois, D. and Prade, H. (1980). *Fuzy Sets and Systems.* Academic Press, New York.

[50] Dubois, D. and Prade, H. (1985). A note on measures of specificity for fuzzy sets. *International Journal of General Systems,* **10**, 279-283.

[51] Dubois, D. and Prade, H. (1986). On the unicity of Dempster's rule of combination. *International Journal of Intelligent Systems,* **1**, 133-142.

[52] Dubois, D. and Prade, H. (1988). *Possibility Theory.* Plenum Press, New York - London.

[53] Dumitriu, A. (1971). *Logica Polivalentă.* Editura Enciclopedică Română, Bucureşti.

[54] Dumitriu, A. (1975). *Istoria Logicii.* (2nd ed.) Editura Didactică şi Pedagogică, Bucureşti.

[55] Dürrenmatt, F. (1958). *Das Versprechen. Requiem auf den Kriminalroman.* Peter Schifferli Verlags AG "Die Arche", Zürich.

[56] Edman, M. (1974). The prediction paradox. *Theoria,* **40**, 166-175.

[57] Edwards, H.M. (1974). *Riemann's Zeta Function.* Academic Press, New York.

[58] Emptoz, H. (1981). Nonprobabilistic entropies and indetermination measures in the setting of fuzzy sets theory. *Fuzzy Sets and Systems,* **5**, 307-317.

[59] Février, P. (1937). Les relations d'incertitude d'Heisenberg et la logique. In *Travaux du IX-ème Congrès International de Philosophie, 1936,* VI, Hermann, Paris, pp.88-94.

[60] Feyerabend, P. (1958). Reichenbach's interpretation of quantum mechanics. *Philosophical Studies,* **9**, 49-59.

[61] Fitch, F.B. (1964). A Goedelized formulation of the prediction paradox. *American Philosophical Quarterly,* **1**, 161-164.

[62] Gaarder, J. (1996). *Sophie's World. A Novel About the History of Philosophy.* Berkley Books, New York.

[63] Gardner, M. (1982). *Logic Machines and Diagrams.* (2nd ed.) University of Chicago Press, Chicago.

[64] Gatlin, L.L. (1972). *Information Theory and the Living System.* Columbia University Press, New York.

[65] Georgescu-Roegen, N. (1971). *The Entropy Law and the Economic Process.* Harvard University Press, Cambridge, Massachusetts.

[66] Gödel, K. (1931). Über Formal Unentscheidbare Sätze der Principia Mathematica und Verwandter Systeme, I. *Monatshefte für Mathematik und Physik,* **38**, 173-198.

[67] Greenstein, C.H. (1978). *Dictionary of Logical Terms and Symbols.* Van Nostrand Reinhold Company, New York.

[68] Gribbin, J. (1984). *In Search of Schrödinger Cat. Quantum Physics and Reality.* Bantam Books, Toronto, New York, London, Sydney, and Auckland.

[69] Guiasu, S. (1971). Weighted entropy. *Reports on Mathematical Physics,* **2**, 165-179.

[70] Guiasu, S. (1977). *Information Theory with Applications.* McGraw-Hill, New York.

[71] Guiasu, S. (1987). Prediction paradox revisited. *Logique et Analyse,* **30**, 147-154.

[72] Guiasu, S. (1993a). A unitary treatment of several known measures of uncertainty induced by probability, possibility, fuzziness, plausibility, and belief. In Bouchon-Meunier, B., Valverde, L., and Yager, R.R. (eds.) *Uncertainty in Intelligent Systems.* North-Holland, Amsterdam-London-New York-Tokyo, pp.355-365.

[73] Guiasu, S. (1993b). Weighting independent bodies of evidence. In Clarke, M., Kruse, R., and Moral, S. (eds.) *Symbolic and Quantitative Approaches to Reasoning and Uncertainty.* Lecture Notes in Computer Science No.747, Springer-Verlag, Berlin-Heidelberg-New York, pp.168-173.

[74] Guiasu, S. (1994a). Fuzzy sets with inner interdependence. *Information Sciences,* **79**, 315-338.

[75] Guiasu, S. (1994b). Reaching a verdict by weighting evidence. In Wang, P.P. (ed.) *Advances in Fuzzy Theory and Technology.* Vol.2. Book-wrights, Raleigh, North Carolina, pp.167-180.

[76] Guiasu, S. (1995a). Stochastic logic. In Bouchon-Meunier, B., Yager, R.R., and Zadeh, L.A. (eds.) *Advances in Intelligent Computing - IPMU'94.* Lecture Notes in Computer Science No.945. Springer-Verlag, Berlin-Heidelberg-New York, pp.370-379.

[77] Guiasu, S. (1995b). On the formalism of stochastic logic. In Froidevaux, C. and Kohlas, J. (eds.) *Symbolic and Quantitative Approaches*

to Reasoning and Uncertainty. Lecture Notes in Artificial Intelligence No.946. Springer-Verlag, Berlin-Heidelberg-New York, pp.227-234.

[78] Guiasu, S. (1999). The process of weighting credibilities. *International Journal of Mathematical and Statistical Sciences*, **8**, 187-197.

[79] Guiasu, S. (2000a). A generalization of Bayes' rule on posterior credibilities. *Proceedings of the Eighth International Conference IPMU*. Vol.3. Consejo Superior de Investigaciones Cientifíficas, Universidad Politécnica de Madrid, pp.1830-1837.

[80] Guiasu, S. (2000b). Coping with uncertainty in n-person games. *International Journal of Uncertainty, Fuzziness, and Knowledge-Based Systems*, **8**, 503-523.

[81] Guiasu, S. (2001). *Quantum Mechanics*. Nova Science Publishers, Inc., Huntington, New York.

[82] Guiasu, S. and Malitza, M. (1980). *Coalition and Connection in Games*. Pergamon Press, Oxford, England.

[83] Hacking, I. (ed.) (1981). *Scientific Revolutions*. Oxford University Press, Oxford.

[84] Hadamard, J. (1954). *An Essay on the Psychology of Invention in the Mathematical Field*. Dover Publications, New York.

[85] Harsanyi, J.C. (1977). *Rational Behaviour and Bargaining Equilibrium in Games and Social Situations*. Cambridge University Press, Cambridge, England.

[86] Hartley, R.V.L. (1928). Transmission of information. *Bell System Technical Journal*, **7**, 535-563.

[87] Hartshorne, C. and Weiss, P. (eds.) (1931-1935). *Pierce, C.S. Collected Papers*. (Volumes 1-6). Harvard University Press, Cambridge, Massachusetts.

[88] Hawking, S.W. (1988). *A Brief History of Time*. Bantam Books, Toronto, New York.

[89] Hay, L.S. (1963). Axiomatization of the infinite-valued predicate calculus. *Journal of Symbolic Logic*, **28**, 77-86.

[90] Hegel, G.W.F. (1843). *Encyklopädie der philosophischen Wissenschaften im Grundrisse. Erster Theil. Die Logik.* Verlag von Dunder, Berlin.

[91] Heisenberg, W. (1927). Über den anschaulichen Inhalt der quantentheoretischen Kinematik und Mechanik. *Zeitschrift für Physik*, **43**, 172-198.

[92] Higashi, M. and Klir, G.J. (1982). Measures of uncertainty and information based on possibility distributions. *International Journal of General Systems*, **9**, 43-58.

[93] Hilbert, D. and Ackermann, W. (1949). *Grundzüge der Theoretischen Logik.* (2nd ed.) Springer-Verlag, Berlin.

[94] Hilbert, D. and Ackermann, W. (1950). *Principles of Mathematical Logic.* Chelsea, New York.

[95] Hirota, H. (1982). Ambiguity based on the concept of subjective entropy. In Gupta, M.M. and Sanchez, E. (eds.) *Fuzzy Information and Decision Processes.* North-Holland Publishing Company, New York, pp.29-40.

[96] Hofstadter, D.R. (1980). *Gödel, Escher, Bach: an Eternal Golden Braid.* Vintage Books (A Division of Random House), New York.

[97] Hofstadter, D.R. (1986). *Metamagical Themas: Questing for the Essence of Mind and Pattern.* Bantam Books, Toronto and New York.

[98] Horibe, Y. (1973). A note on entropy metrics. *Information and Control*, **22**, 403-404.

[99] Horsley, E.M. (ed.) (1981). *Hutschinson 20th Century Encyclopedia.* (7th ed.) Hutchinson & Co. Publishers, London.

[100] Hybel, A.R. (1986). *The Logic of Surprise in International Conflict.* Lexington Books, D.C. Heath and Company, Lexington, Massachusetts & Toronto.

[101] Ichihashi, H. and Tanaka, H. (1989). Jeffrey-like rules of conditioning for the Dempster-Shafer theory of evidence. *International Journal of Approximate Reasoning*, **3**, 143-156.

[102] Jacob, F. (1982). *The Logic of Life*. Pantheon Books, New York.

[103] Jain, L.C., Howlett, R.J., Ichalkaranje, N.S., and Tonfoni, G. (eds.) (2002). *Virtual Environments for Teaching & Learning*. World Scientific, New Jersey, London, Singapore, and Hong Kong.

[104] Jantsch, E. (ed.) (1981). *The Evolutionary Vision*. Westview Press, Boulder, Colorado.

[105] Jaynes, E.T. (1979). Where do we stand on maximum entropy? In Levin, R. and Tribus, M. (eds.) *The Maximum Entropy Formalism*. MIT Press, Cambridge, Massachusetts, pp.15-118.

[106] Jeffrey, R.C. (1965). *The Logic of Decision*. McGraw-Hill, New York.

[107] Jumarie, G. (1990). A theory of information for vague concepts. Outline of application to approximate reasoning. *Kibernetes*, **19**, 15-34.

[108] Kam, E. (1988). *Surprise Attack. The Victim's Perspective*. Harvard University Press, Cambridge, Massachusetts and London, England.

[109] Keynes, J.M. (1921). *A Treatise on Probability*. Macmillan, London.

[110] Klir, G.J. (1987). Where do we stand on measures of uncertainty, ambiguity, fuzziness and the like? *Fuzzy Sets and Systems*, **24**, 141-160.

[111] Kolmogorov, A.N. (1950). *Foundations of the Theory of Probability*. Chelsea, New York.

[112] Kyburg, H. (1961). *Probability and the Logic of Rational Belief*. Wesleyan University Press, Middletown, Connecticut.

[113] Lamata, M.T. and Moral, S. (1988). Measures of entropy in the theory of evidence. *International Journal of General Systems*, **14**, 297-305.

[114] Landa, L.N. (1962). Logical-informational algorithm for learning theory. *Psychological Journal* (in Russian), **2**, 19-40.

[115] Laplace, P.S. (1812). *Thórie analytique des probabilités*. Courcier, Paris.

[116] Levi, I. (1973). *Gambling with Truth*. The MIT Press, Cambridge, Massachusetts and London, England.

[117] Lewis, C.I. (1918). *A Survey of Symbolic Logic*. University of California Press, Berkeley.

[118] Lewis, C.I. and Langford, C.H. (1956). History of symbolic logic. In Newman, J.R. (ed.) *The World of Mathematics*. Vol.3 (11th edition), Simon and Schuster, New York, pp.1859-1877.

[119] Lewis, P.M. (1962). The characteristic selection problem in recognition systems. *IRE Transactions on Information Theory*, **IT-8**, 171-178.

[120] Lindley, D.V. (1987a). The probability approach to the treatment of uncertainty in artificial intelligence and expert systems. *Statistical Sciences*, **2**, 17-24.

[121] Lindley, D.V. (1987b). Comment: A tale of two wells. *Statistical Sciences*, **2**, 38-40.

[122] Luce, R.D. and Raiffa, H. (1967). *Games and Decisions*. Wiley and Sons, New York.

[123] Lukasiewicz, J. (1930). Philosophische Bemerkungen zur mehrwertigen Systemen des Aussagenkalküls. *Comptes Rendus des Séances de la Société des Sciences et des Lettres de Varsovie*, **23**, Classe III, 51-77.

[124] Lukasiewicz, J. (1957). *Aristotle's Syllogistic from the Standpoint of Modern Formal Logic*. Clarendon Press, Oxford.

[125] Lukasiewicz, J. and Tarski, A. (1930). Unterschungen über den Aussagenkalkül. *Comptes Rendus des Séances de la Société des Sciences et des Lettres de Varsovie*, **23**, Classe III, 30-50.

[126] MacColl, H. (1906). *Symbolic Logic and Its Applications*. Longmans, London.

[127] Maida, P.D. and Spornick, N.B. (1982). *Murder She Wrote.* Bowling Green University Popular Press, Ohio.

[128] Markov, A.A. (1906). Extension of the law of large numbers to dependent events. (Russian). *Bulletin de la Societé de Physique et Mathématique de Kazan (2)*, **15**, 135-156.

[129] Martin, R.M., (1992). *There Are Two Errors in the the Title of This Book.* Broadview Press, Peterborough, Ontario.

[130] Medlin, B. (1964). The unexpected examination. *American Philosophical Quarterly*, **1**, 66-72.

[131] Mill, J.St. (1896). *Sistème de logique deductive et inductive.* Félix Alcan, Paris.

[132] Mill, J.St. (1980). *On Liberty.* Bobbs-Merrill Educational Publishing, Indianopolis.

[133] Monod, J. (1974). *Chance and Necessity.* Collins, Fontana Books, Glasgow.

[134] Morris, R. (1984). *Time's Arrows.* Simon and Schuster, New York.

[135] Nagel, E. (1956). Symbolic notation, Haddocks' eyes and the dog-walking ordinance. In Newman, J.R. (ed.) *The World of Mathematics.* Vol.3 (11th edition), Simon and Schuster, New York, pp.1878-1900.

[136] Nagel, E. and Newman, J.R. (1958). *Gödel's Proof.* New York University Press, New York.

[137] Negoita, C.V. and Ralescu, D.A. (1975). *Applications of Fuzzy Sets to System Analysis.* Birkhauser, Stuttgart.

[138] Nguyen, H.T. (1985). On entropy of random sets and possibility distributions. In Bezdek, J.C. (ed.) *The Analysis of Fuzzy Information.* CRC Press, Boca Raton, Florida.

[139] Nicod, J. (1916). A reduction in the number of the primitive propositions of logic. *Proceedings of the Cambridge Philosophical Society*, **19**, 32-42.

[140] Olin, D. (1983). The prediction paradox resolved. *Philosophical Studies*, **44**, 225-233.

[141] Osborne, C.(1982). *The Life and Crimes of Agatha Christie*. Collins, London.

[142] Park, D. (1980). *The Image of Eternity. Roots of Time in the Physical World*. The University of Massachusetts Press, Amherst.

[143] Parkinson, G.H.R. (ed.) (1981). *Leibniz. Philosophical Writings*. Rowman and Littlefield, Totowa, New Jersey.

[144] Paulos, J.A. (1980). *Mathematics and Humor*. The University of Chicago Press, Chicago and London.

[145] Paulos, J.A. (1985). *I Think, Therefore I Laugh*. Columbia University Press, New York.

[146] Paulos, J.A.,(1991). *Beyond Numeracy*. Alfred A. Knopf, New York.

[147] Pawlak, Z., Wong, S.K.M., and Ziarko, M. (1988). Rough sets: probabilistic versus deterministic approach. *International Journal of Man-Machine Studies*, **29**, 81-95.

[148] Pearson, K. (1900). On the criterion that a given system of deviations from the probable in the case of a correlated system of variables is such that it can be reasonably supposed to have arisen from random sampling. *Philosophical Magazine* (5th series), **50**, 157-175.

[149] Peirce, C.S. (1880). On the algebra of logic. *American Journal of Mathematics*, **3**, 15-57.

[150] Penrose, R. (1989). *The Emperor's New Mind*. Oxford University Press, Oxford, U.K.

[151] Péter, R. (1961). *Playing with Infinity*. Dover Publications, New York.

[152] Philipps, L. (1964). Rechtiche Regelung und formale Logik. *Archiv für Rechts und Sozialphilosophie*, **50**, 317-329.

[153] Philipps, L. (1966). Sinn und Struktur der Norm Logik. *Archiv für Rechts und Sozialphilosophie*, **52**, 195-219.

[154] Poincaré, H. (1913). *The Foundations of Science.* The Science Press, New York.

[155] Polya, G. (1954). *Mathematics and Plausible Reasoning.* Princeton University Press, Princeton, New Jersey.

[156] Popper, K.R. (1968a). *The Logic of Scientific Discovery.* (2nd ed.) Harper and Row, New York, Hagerstown, San Francisco, and London.

[157] Popper, K.R. (1968b). *Conjectures and Refutations: The Growth of Scientific Knowledge.* Harper and Row, New York, Hagerstown, San Francisco, and London.

[158] Popper, K.R. (1981). *Objective Knowledge. An Evolutionary Approach.* Oxford University Press, Oxford.

[159] Post, E.L. (1921). Introduction to a general theory of elementary propositions. *American Journal of Mathematics*, **40**, 163-183.

[160] Powers, J. (1982). *Philosophy and the New Physics.* Methuen, London and New York.

[161] Prior, A.N. (1955). Many-valued and modal systems: an intuitive approach. *Philosophical Review*, **60**, 626-630.

[162] Putnam, H. (1957). Three-valued logic. *Philosophical Sudies*, **8**, 73-80.

[163] Quine, W.V. (1953). On a supposed paradox. *Mind*, **62**, 65-67.

[164] Rapoport, A. (2001). *N-Person Game Theory. Concepts and Applications.* Dover Publications, Mineola, New York.

[165] Rapoport, A. and Chammah, A.M. (1965). *Prisoner's Dilemma. Conflict and Cooperation.* University of Michigan Press, Ann Arbor, Michigan.

[166] Reghis, M. and Roventa, E. (1998). *Classical and Fuzzy Concepts in Mathematical Logic and Applications.* CRC Press, Boca Raton, Florida.

[167] Reichenbach, H. (1944). *Philosophical Foundations of Quantum Mechanics*. Berkeley, Los Angeles.

[168] Reichenbach, H. (1949). *The Theory of Probability*. Berkeley, Los Angeles.

[169] Rényi, A. (1955). On a new axiomatic theory of probability. *Acta Mathematica of the Academy of Sciences of Hungary*, **6**, 285-335.

[170] Rényi, A. (1972). *Letters on Probability*. Wayne State University Press, Detroit.

[171] Rescher, N. (1966). *The Logic of Command*. Routledge and Kegan Paul Ltd., London, and Dover Publications Inc., New York.

[172] Riley, D. and McAllister, P. (eds.) (1979). *The New Bedside, Bathtub and Armchair Companion to Agatha Christie*. Ungar Publishing Co., New York.

[173] Russell, B. (1903). *The Principles of Mathematics*. Cambridge University Press, Cambridge, England, & New York.

[174] Sainsbury, R.M. (1988). *Paradoxes*. Cambridge University Press, Cambridge, England, & New York.

[175] Sander, W. (1989). On measure of fuzziness. *Fuzzy Sets and Systems*, **29**, 49-55.

[176] Savage, L.J. (1954). *The Foundations of Statistics*. Wiley and Sons, New York.

[177] Scriven, M. (1951). Paradoxical announcements. *Mind*, **60**, 403-407.

[178] Shafer, G. (1976). *A Mathematical Theory of Evidence*. Princeton University Press, Princeton, New Jersey.

[179] Shafer, G. (1990). Perspectives on the theory and practice of belief functions. *International Journal of Approximate Reasoning*, **4**, 323-362.

[180] Shannon, C.E. (1948). A mathematical theory of communication. *Bell System Technical Journal*, **27**, 379-423, 623-656.

[181] Shapley, L.S. (1953). A value for *n*-person games. In Kuhn, H. and Tucker, A. (eds.) *Contributions to the Theory of Games.* Vol.2. Princeton University Press, Princeton, New Jersey, pp.307-317.

[182] Shaw-Kwei, M. (1954). Logical paradoxes for many-valued systems. *Journal of Symbolic Logic*, **19**, 37-39.

[183] Smith, J.A. and Ross, W.D. (eds.) (1908-1931). *The Oxford Translation of Aristotle.* (Volumes 1-11). Oxford University Press, Oxford.

[184] Smullyan, R.M. (1992). *Gödel's Incompleteness Theorems.* Oxford University Press, New York and Oxford.

[185] Tammelo, I. (1964). Law, logic, and human communication. *Archiv für Rechts und Sozialphilosophie*, **50**, 331-366.

[186] Tankard Jr., J.W. (1984). *The Statistical Pioneers.* Schenkman Publishing Company, Cambridge, Massachusetts.

[187] Tarski, A. (1956). Symbolic logic. In Newman, J.R. (ed.) *The World of Mathematics.* Vol.3 (11th edition), Simon and Schuster, New York, pp.1901-1931.

[188] Thomas, L.C. (1984). *Games. Theory and Applications.* Ellis Horwood Ltd., Chichester.

[189] Toulmin, S. and Goodfield, J. (1982). *The Discovery of Time.* The University of Chicago Press, Chicago and London.

[190] Tzara, T. (1984). *Seven Dada Manifestos and Lampisteries.* John Calder, London; Riverrun Press, New York.

[191] Venn, J. (1876). Boole's logical system. *Mind*, **1**, 479-491.

[192] Von Mises, R. (1956). Mathematical postulates and human understanding. In Newman, J.R. (ed.) *The World of Mathematics.* Vol.3 (11th edition), Simon and Schuster, New York, pp.1723-1754.

[193] Von Neumann, J. and Morgenstern, O. (1944). *Theory of Games and Economic Behavior.* Princeton University Press, Princeton, New Jersey.

[194] Wald, A. (1960). *Statistical Decision Functions.* Wiley and Sons, New York.

[195] Watanabe, S. (1969). *Knowing and Guessing. A Quantitative Study of Inference and Information.* Academic Press, New York.

[196] Weinberger, O. (1957). Über die Negation von Sollsätzen. *Theoria,* **23**, 102-132.

[197] Weintraub, R. (1995). Practical solutions to the surprise-examination paradox. *Ratio* **7**, 160-169.

[198] Whitehead, A.N. (1967). *Adventures of Ideas.* The Free Press, New York.

[199] Whitehead, A.N. and Russell, B. (1925). *Principia Mathematica.* Cambridge University Press, Cambridge, England.

[200] Wilder, R.L. (1956). The axiomatic method. In Newman, J.R. (ed.) *The World of Mathematics.* Vol.3 (11th edition), Simon and Schuster, New York, pp.1647-1667.

[201] Wittgenstein, L. (1933). *Tractatus Logico-Philosophicus.* Ed. Paul Kegan , London.

[202] Yager, R.R. (1979). On the measure of fuziness and negation. Part I: Membership in the unit interval. *International Journal of General Systems,* **5**, 221-229.

[203] Yager, R.R. (1983). Entropy and specificity in a mathematical theory of evidence. *International Journal of General Systems,* **9**, 249-260.

[204] Yeh, J. (1973). *Stochastic Processes and the Wiener Integral.* Marcel Dekker Inc., New York.

[205] Zadeh, L.A. (1965). Fuzzy sets. *Information and Control,* **1**, 338-353.

[206] Zadeh, L.A. (1968). Probability measure of fuzzy events. *Journal of Mathematical Analysis and Applications,* **23**, 421-427.

[207] Zadeh, L.A. (1978). Fuzzy sets as a basis for a theory of possibility. *International Journal of Fuzzy Sets and Systems,* **1**, 3-28.

Index